Physics

MYP by Concept
4 & 5

Physics

Paul Morris

Author's acknowledgements

My eternal thanks to Louise Harrison, for her support in taking on more than her fair share of parenting while I wrote this book.

I would like to dedicate this book to my father, Dr Christopher Morris, who started it all when he let me play with the contents of his electronic components box.

I would like to thank the following for their advice and support in the development of this project: Rita Bateson, Curriculum and Assessment Manager for Mathematics and Sciences, Middle Years Programme, IBO; Robert Harrison, Head of Middle Years Programme Development, IBO; Lennox Meldrum, International School KL, and So-Shan Au, International Publisher at Hodder Education, for her patience and imagination.

Orders: please contact Hachette UK Distribution, Hely Hutchinson Centre, Milton Road, Didcot, Oxfordshire, OX11 7HH. Telephone: +44 (0)1235 827827. Email education@hachette.co.uk Lines are open from 9 a.m. to 5 p.m., Monday to Friday. You can also order through our website: www.hoddereducation.com

First published in 2015 by
Hodder Education,
An Hachette UK Company
Carmelite House
50 Victoria Embankment
London EC4Y 0DZ

Impression number   10 9 8 7
Year   2021

Cover photo © alexskopje – Fotolia
Illustrations by Barking Dog Art and DC Graphic Design Limited
Typeset in Frutiger LT Std 45 Light 10/14pt by DC Graphic Design Limited, Hextable, Kent
Printed in Dubai

A catalogue record for this title is available from the British Library.

ISBN 9781471839337

Contents

How to use this book

Welcome to Hodder Education's *MYP by Concept* series! Each chapter is designed to lead you through an *inquiry* into the concepts of physics, and how they interact in real-life global contexts.

Each chapter is framed with a *Key concept* and a *Related concept* and is set in a *Global context*.

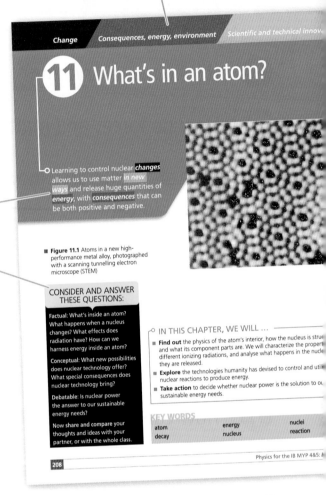

| Change | Consequences, energy, environment | Scientific and technical innova |

11 What's in an atom?

Learning to control nuclear *changes* allows us to use matter *in new ways* and release huge quantities of *energy*, with *consequences* that can be both positive and negative.

■ **Figure 11.1** Atoms in a new high-performance metal alloy, photographed with a scanning tunnelling electron microscope (STEM)

CONSIDER AND ANSWER THESE QUESTIONS:

Factual: What's inside an atom? What happens when a nucleus changes? What effects does radiation have? How can we harness energy inside an atom?

Conceptual: What new possibilities does nuclear technology offer? What special consequences does nuclear technology bring?

Debatable: Is nuclear power the answer to our sustainable energy needs?

Now share and compare your thoughts and ideas with your partner, or with the whole class.

IN THIS CHAPTER, WE WILL ...

■ **Find out** the physics of the atom's interior, how the nucleus is stru and what its component parts are. We will characterize the propert different ionizing radiations, and analyse what happens in the nucle they are released.

■ **Explore** the technologies humanity has devised to control and utili nuclear reactions to produce energy.

■ **Take action** to decide whether nuclear power is the solution to ou sustainable energy needs.

KEY WORDS

| atom | energy | nuclei |
| decay | nucleus | reaction |

Physics for the IB MYP 4&5:

208

KEY WORDS

Key words are included to give you access to vocabulary for the topic. **Glossary** terms are highlighted and given where applicable. Search terms are given to encourage independent learning and research skills.

As you explore, activities suggest ways to learn through *action*.

The *Statement of Inquiry* provides the framework for this inquiry, and the *Inquiry questions* then lead us through the exploration as they are developed through each chapter.

■ ATL

Activities are designed to develop your *Approaches to Learning (ATL)* skills.

◆ Assessment opportunities in this chapter

Some activities are *formative* as they allow you to practice skills that are assessed using MYP Sciences assessment criteria. Other activities can be used by you or your teachers to assess your achievement *summatively* against all parts of a learning objective.

Key *Approaches to Learning* skills for MYP Sciences are highlighted whenever we encounter them.

Hint

In some of the activities, we provide hints to help you work on the assignment. This also introduces you to the new hint feature in the on-screen assessment.

EXTENSION

Extension activities allow you to explore a topic further.

You can measure your conceptual understanding using the summary problems at the end of every chapter, organized by level of difficulty.

Finally, at the end of the chapter you are asked to reflect back on what you have learned with our *Reflection table*, and maybe to think of new questions brought to light by your learning.

Reflecting on our learning... Use this table to reflect on your own learning in this chapter.					
Questions we asked	**Answers we found**	**Any further questions now?**			
Factual					
Conceptual					
Debatable					
Approaches to learning you used in this chapter:	**Description – what new skills did you learn?**	**How well did you master the skills?**			
		Novice	Learner	Practitioner	Expert
Collaboration skills					
Information Literacy					
Critical thinking skills					
Creative thinking skills					
Learner profile attribute(s)	*Reflect on the importance of the attribute for our learning in this chapter.*				
Knowledgeable					

We have incorporated Visible Thinking – ideas, framework, protocol and thinking routines – from Project Zero at the Harvard Graduate School of Education into many of our activities.

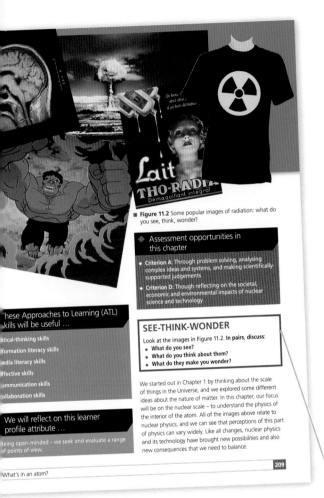

■ **Figure 11.2** Some popular images of radiation: what do you see, think, wonder?

◆ Assessment opportunities in this chapter

◆ **Criterion A:** Through problem solving, analysing complex ideas and systems, and making scientifically-supported judgements
◆ **Criterion D:** Though reflecting on the societal, economic and environmental impacts of nuclear science and technology

These Approaches to Learning (ATL) skills will be useful ...

itical-thinking skills
formation literacy skills
edia literacy skills
ffective skills
mmunication skills
ollaboration skills

We will reflect on this learner profile attribute ...

eing open-minded – we seek and evaluate a range f points of view.

What's in an atom?

SEE-THINK-WONDER

Look at the images in Figure 11.2. **In pairs, discuss:**
• What do you see?
• What do you think about them?
• What do they make you wonder?

We started out in Chapter 1 by thinking about the scale of things in the Universe, and we explored some different ideas about the nature of matter. In this chapter, our focus will be on the nuclear scale – to understand the physics of the interior of the atom. All of the images above relate to nuclear physics, and we can see that perceptions of this part of physics can vary widely. Like all changes, nuclear physics and its technology have brought new possibilities and also new consequences that we need to balance.

209

Take action

! **While the book provides opportunities for action and plenty of content to enrich the conceptual relationships, you must be an active part of this process. Guidance is given to help you with your own research, including how to carry out research, guidance on forming your own research question, as well as linking and developing your study of Physics to the global issues in our twenty-first century world.**

▼ Links

Like any other subject, Physics is just one part of our bigger picture of the world. Links to other subjects are discussed.

● Learner profile attribute

● Each chapter has an **IB Learner Profile** attribute as its theme, and you are encouraged to reflect on these too.

① How big is everything?

○ We understand our own **relationship** to the Universe when we identify *patterns* at *different scales*.

■ **Figure 1.1** How does changing scale change the way the Universe looks?

CONSIDER AND ANSWER THESE QUESTIONS:

Factual: What is the smallest thing? What is the largest thing? How do we measure them? What is the Universe made from?

Conceptual: How is the Universe structured? How are the smallest and the largest things in the Universe connected?

Debatable: What would the world look like if you were very small? Or very large?

Now **share and compare** your thoughts and ideas with your partner, or with the whole class.

○ IN THIS CHAPTER WE WILL …

■ **Find out** how the Universe is structured, from the very smallest observable sizes to the very largest.

■ **Explore** the various ideas that humanity has held at different times about the nature of the 'stuff' in the Universe, and how different patterns at the smallest of scales can make the biggest differences.

■ **Take action** to research how new materials might be able to help those in less economically developed parts of the world.

● We will reflect on this learner profile attribute …

● Inquirers – we will explore and research new scales of observation, and new materials.

■ These Approaches to Learning (ATL) skills will be useful …

■ **Critical-thinking skills**

■ **Creative-thinking skills**

■ **Transfer skills**

■ **Information literacy skills**

■ **Communication skills**

THE SCALE OF THINGS

We refer to the size of the objects we can observe as their **scale**.

Some time ago two artists – Ray and Charles Eames – made an animated film about the idea of scale. It was called *Powers of Ten*. To find out about the original film and to watch it, use YouTube **Powers of 10**. A number of websites have used and updated the Eames' idea.

◆ Assessment opportunities in this chapter

◆ **Criterion A**: Knowledge and understanding

◆ **Criterion B**: Inquiring and designing

◆ **Criterion D**: Reflecting on the impacts of science

ACTIVITY: It depends on your point of view …

■ ATL

■ **Information literacy skills**: Understand and use technology systems

● **Explore** online using some of these links to find out what the Universe looks like at different scales (apps may require Java to be installed on your computer):

http://apod.nasa.gov/apod/ap120312.html
http://micro.magnet.fsu.edu/primer/java/scienceopticsu/powersof10/

● **Find out** what the Universe 'looks like':

o at the **smallest** known scales. What are the smallest objects?
o at the **largest** known scales. What are the largest objects?

What would the world look like if you were very small? Or very large?

OBSERVATIONS AT DIFFERENT SCALES

Science is all about explaining observations, so scientists have spent a lot of time and effort inventing and improving devices for making observations at different scales. One of the most significant 'leaps' of understanding came with the discovery that glass could be ground and shaped to focus light from distant objects, such that they formed images that appeared larger than the objects themselves, making more detail available to the human eye. An early lens was found dating back to 700 BCE, and lenses are mentioned in plays from ancient Greece written around 400 BCE. By the 13th century CE spectacles were in common use across Europe.

In the early 17th century CE a Dutch spectacle maker had the idea of placing two lenses together, such that the second lens magnified the light gathered by the first. Shortly afterwards, the Italian philosopher and mathematician Galileo Galilei made his own version of the **telescope** and, for the first time, observed the four largest moons of Jupiter in orbit around the planet.

▼ Link: Mathematics

■ ATL

■ **Transfer skills**: Make connections between subject groups and disciplines

Important key concepts in MYP Mathematics are **form** and **relationships**:
- **Form in mathematics** concerns the way that appearances can tell us about purpose or the function – what something does.

- **Relationships in mathematics** concerns the way that things connect together.

So, it should not surprise us that mathematics can help us work with relationships between things at different scales! When scientists are dealing with very large scales or very small scales, they have to calculate very large and very small numbers. Our usual unit of length – the metre – is made to be useful to us, in our 'scale' for the Universe. The movie *Powers of Ten* demonstrates how **standard notation** is used to help us represent size.

In standard notation:
- 10^1 is said 'ten to the power of one' $= 10$
- 10^2 is said 'ten to the power of two' $= 10 \times 10$
  $= 100$
- 10^3 is said 'ten to the power of three' $= 10 \times 10 \times 10$
  $= 1000$

and so on …

For small numbers we have to go the other way, making the power of ten smaller and smaller. We have to use negative powers of ten to do this:

- 10^{-1} is said 'ten to the power of minus one'
  $= 1 \div 10$ $= \dfrac{1}{10}$ $= 0.1$
- 10^{-2} is said 'ten to the power of minus two'
  $= 1 \div (10 \times 10)$ $= \dfrac{1}{100}$ $= 0.01$
- 10^{-3} is said 'ten to the power of minus three'
  $= 1 \div (10 \times 10 \times 10)$ $= \dfrac{1}{1000}$ $= 0.001$

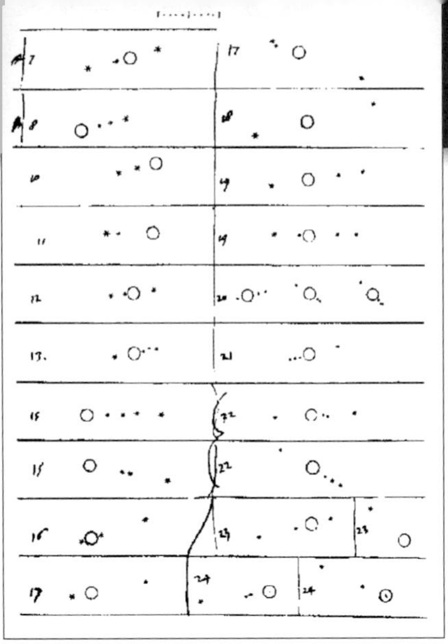

Figure 1.2 Galileo's sketches of the four principal moons of Jupiter, now named the 'Galilean moons' in his honour

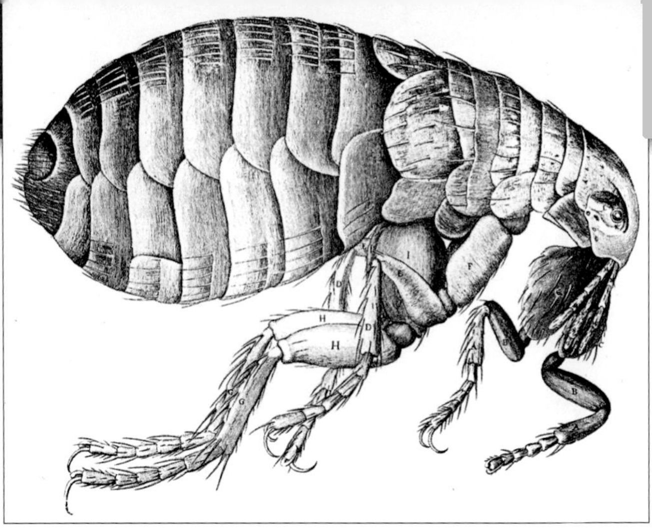

Figure 1.3 Drawing of a flea from Robert Hooke's *Micrographia* (1665)

Equally, the same system of lenses could be used to observe very small objects that were very close to the lens. The first **microscope** is attributed to Zacharias Jansen in the 1590s. Shortly afterwards, in 1665, the first systematic observations using a microscope were published by the English 'natural philosopher' Robert Hooke.

Observations have benefitted from improvements in **magnification** and **resolution** ever since. The magnification of an instrument is the increase in apparent size of an image relative to the object's size, while the

You can browse Hooke's drawings in the e-book at Project Gutenburg: **www.gutenberg.org**

resolution is the smallest distance between two points that can be distinguished using the instrument.

While we now have instruments that work with information other than that carried by visible light, the aim of all instruments is the same: to make that which is invisible to the human eye available to our understanding (Table 1.1).

Table 1.1

Observation instruments	Magnification (how much bigger the image is than the object)	Resolution (smallest distance clearly visible)
Hubble space telescope	× 8000 (estimated)	0.05 arcseconds
magnifying glass	× 2	10^{-5} m
scanning electron microscope	× 500 000	10^{-10} m
laboratory microscope	× 1500	10^{-7} m
astronomical telescope	× 300	2 arcseconds
human eye	× 1	0.5×10^{-3} m

EXTENSION

Explore further! Did you notice that the **resolution** for telescopes is given as 'arcseconds', while for microscopes it is given as a distance in metres? Why do you think this might be? For a clue, read about telescope resolution here:

http://hubblesite.org/the_telescope/

Select Nuts and bolts then resolution 101.

See Chapter 12 for more on optical instruments and observations of space.

What is the largest thing? How do we measure them?

A RULER FOR THE UNIVERSE

Since scientists have to use large and small numbers so often, it would be quite tedious to write out the dimensions of objects in metres all the time, along with the necessary powers of ten. For this reason, the *Système International d'Unités* or S.I. unit system includes a series of prefixes for different scales. A new prefix is used every time a multiple of 1000 or 10^3 occurs. You may have encountered some of these already in relation to computer systems (Table 1.2).

■ Table 1.2

Power of ten	Prefix
× 10^{-15}	femi
× 10^{-12}	pico
× 10^{-9}	nano
× 10^{-6}	micro
× 10^{-3}	milli
× 10^{-1}	deci
× 10^3	kilo
× 10^6	mega
× 10^9	giga
× 10^{12}	tera

WHAT IS THE UNIVERSE MADE FROM?

Humans all over the world, at many different times, have tried to figure out what makes everything. What is the Universe of 'stuff' we see around us? What makes it the way it is? Even before we had instruments that enable us to see down to the tiniest scales, people had ideas about it.

A society that gave this some consideration was a civilization located around the Mediterranean sea – in the countries we now call Greece, Turkey, and the Balkans. We call these people the Greeks, because the only written records that we have come largely from that region, especially from the city of Athens (in fact, the Greek Empire, at one stage, went as far east as India and as far south as Egypt).

ACTIVITY: A ruler for the Universe

■ ATL

- **Information literacy skills**: Access information to be informed and inform others
- **Communication skills**: Use a variety of media to communicate with a range of audiences

Taking inspiration from the *Powers of Ten* film and other websites that you researched, **make a large display, model or diorama** to show others the scale of the known Universe.

▼ Links: Communication and design

How will you communicate the idea of different **scales** using your model? **Research** and **suggest** ways in which your design will do this. For example, if you have a museum or science exhibition nearby, look at the ways in which the exhibits **communicate** ideas in science in different ways:

- www.sciencemuseum.org.uk/
- www.huffingtonpost.com/2013/07/08/ best-science-museums-in-the-us-ranked- mensa_n_3536208.html

◆ Assessment opportunities

In this task you have practised skills that can be assessed using Criterion D: Reflecting on the impacts of science.

- ◆ apply scientific language effectively
- ◆ document the work of others and sources of information used.

(Note, however, that the level you achieve will not represent your overall level of achievement in Criterion D, because there are another two strands to complete!)

Greek thinkers were interested to find out whether the Universe had an 'essential' or 'basic' nature – and, if it did, whether it was possible for us to know anything about that fundamental nature. Was there a 'stuff' that makes up everything? Two Greek thinkers had two very different ideas about this. Read their thoughts in Figures 1.4 and 1.5.

If you ask me, it makes no sense to think of 'nothing'. There must ALWAYS be something – you can't just have an empty space. So, I think that, if you cut matter up, you just get smaller and smaller pieces, forever. **Matter is continuous** and everything is full of it.

Nothing can go on forever, can it? All things must begin and all things must end. So it is with matter – if we cut matter over and over, we will eventually come to a **tiny, basic piece** that is uncuttable. All things, in the end, are made from these uncuttable objects – in my language, the word for this is '**atomos**'.

■ **Figure 1.4** Parmenides
(515–540 BCE)

■ **Figure 1.5** Democritus
(460–370 BCE)

ACTIVITY: Thinking about stuff

Think-pair-share the theories of Parmenides and Democritus.

■ ATL

■ **Critical-thinking skills**: Recognize and evaluate propositions; Evaluate evidence and arguments; Consider ideas from multiple perspectives

Compare the theories to what we learned earlier about the nature of the Universe at the smallest scales known. Think about how matter 'looks' at different scales, and consider whether it looks 'continuous' (as Parmenides thought) or 'made of pieces' (as Democritus thought).

Organize your ideas using a table like the one below.

Evidence that might support Parmenides' idea	Evidence that might support Democritus' idea

What you think

A key skill in science is to decide on the **quality** of the evidence available to support an idea. The quality of scientific evidence is often measured in terms of its validity and its reliability. Use search terms **validity** and **reliability** to find out what these words mean.

How is the Universe structured?

The way the Universe looks depends on the scale we use. At each scale, we see that the 'stuff' that makes things up is itself made from smaller and smaller bits of 'stuff'. This is one pattern we can be sure about! Nobody is really sure whether Parmenides or Democritus was correct; scientific theories about the nature of matter have changed from one view to the other over the two millennia since! Of course, we have a big advantage over Parmenides and Democritus: that advantage is called science. To Greek classical thinkers, the world of the senses – that which we observe – was unreliable and uncertain, because the senses could be deceived. For many of them (with notable exceptions, such as Aristotle) true knowledge could only be obtained through thinking. This view of knowledge is sometimes called **rationalism**.

In Europe, it took a long time to change this view, and a large number of writers and thinkers contributed to the process, such as Thomas Aquinas, Francis Bacon and Galileo Galilei. By around 1600, however, European thinkers were beginning to reconsider the role of observation in making knowledge. Francis Bacon, for example, asked: why should thinking be any more reliable or certain than what we, ourselves, can experience? He proposed that we could be more certain about the world if we made careful, controlled observations of the way things behaved in it.

The idea that knowledge can be gained from observation and experience is sometimes called **empiricism**.

However, if our knowledge of the world is to be reliable, not just any old experience or observation will do. Our observations must be carefully controlled so that we can determine the exact relationships between things – in other words, we must do an **experiment**.

■ **Figure 1.6** Sir Francis Bacon, 1561–1626

■ **Figure 1.7** Galileo Galilei, 1564–1642

▼ Link: The language of science

■ ATL

■ **Transfer skills**: Inquire in different contexts to gain a different perspective

The word **experiment** derives from the Latin *experiri*, meaning **experience** – if you study a Latinate language such as Italian, French or Spanish, you may know that the word for 'experiment' is exactly the same as the word for 'experience'.

EXTENSION

Explore further! Use Google 'ngrams' to find out when the word 'experiment' began to be used more frequently in books and other texts:

https://books.google.com/ngrams/

Galileo Galilei was one of the first people to realize that reliable experiments required controlled variables. A **variable** is any factor that can be controlled or measured in order to experimentally investigate a relationship.

ACTIVITY: Thinking about experimental inquiry

Imagine an experimental inquiry to be like a sort of machine. The controls of the machine determine the outcomes that we can measure.

Here is an example:

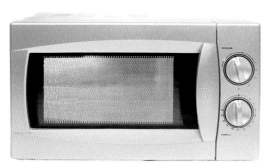

How quickly can my microwave oven heat a hot lunch

Time to heat lunch

Final temperature of lunch

■ **Figure 1.8** Experimental inquiry question   ■ **Figure 1.9** Controlled   ■ **Figure 1.10** Measured outcome

THINK-PAIR-SHARE

What responses do you have to these questions? Think about them on your own, then share with your partner. Then discuss as a class to achieve a class consensus.

- **What other variables might affect the temperature of the lunch?**
- **How can the experiment be designed so that we can account for the effects of these other variables?**

If we change any of these, they might also affect the outcome – the final temperature of my lunch.

We, therefore, have to divide the **controlled** variables into those we will **change**, and those we will **keep the same**; we can modify our diagram accordingly.

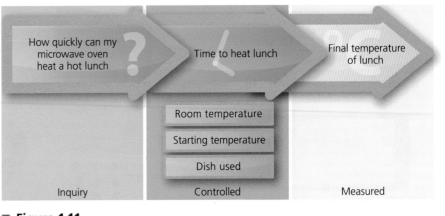

■ **Figure 1.11**

ACTIVITY: The right variable for the job

For the different experimental inquiries listed, choose variables from the box that might be important to **control by changing, control by keeping the same** or to **measure**.

Variables			
temperature	time	height	colour
brightness	distance	area	speed

Experimental inquiries:

- **How long does an ice cube stay frozen?**
- **What size of parachute will save a dropped egg?**
- **What is the best colour to wear if you want to keep cool?**
- **How long does an aircraft stay on the runway before it can take off?**

EXTENSION

Explore further! Can you think of any other experimental inquiries you could do using these variables?

How are the smallest and the largest things in the Universe connected?

MAPPING MATTER

We have improved our knowledge of matter a great deal since Parmenides' and Democritus' time. Today we know of the existence of atoms with some certainty. We also know that atoms themselves consist of smaller parts, and it is generally held by scientists that some of those parts (the nucleons, meaning the protons and the neutrons) consist of smaller parts called **quarks** … and then what?

The French aristocrat Antoine Lavoisier was one of the first people to suggest that all matter is made from only a certain number of 'building blocks', or **elements** – fundamental substances that consist of atoms of only one kind and that, singly or in combination, constitute all matter. In the late 18th century CE he managed to prove that water was made from **hydrogen** and **oxygen**. He also wrote one of the first lists of elements.

■ **Figure 1.13** Individual atoms seen through a scanning tunnelling electron microscope

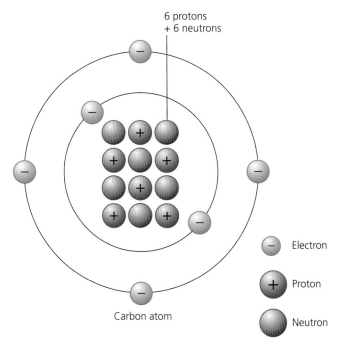

6 protons
+ 6 neutrons

Carbon atom

Electron

Proton

Neutron

■ **Figure 1.12** A carbon atom

■ **Figure 1.14** Portrait of Lavoisier and his wife, By Jacques Louis David (Musée du Louvre, Paris)

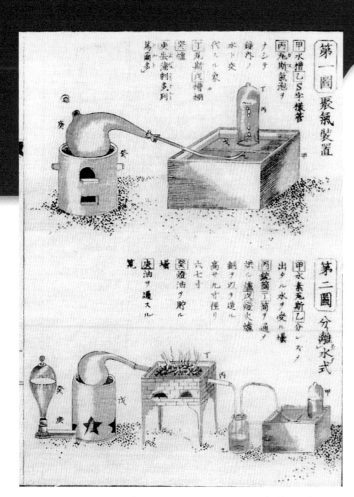

■ **Figure 1.15** A 19th century Japanese translation of Lavoisier's experiments

■ **Figure 1.16** Dimitri Mendeleev (1834–1907)

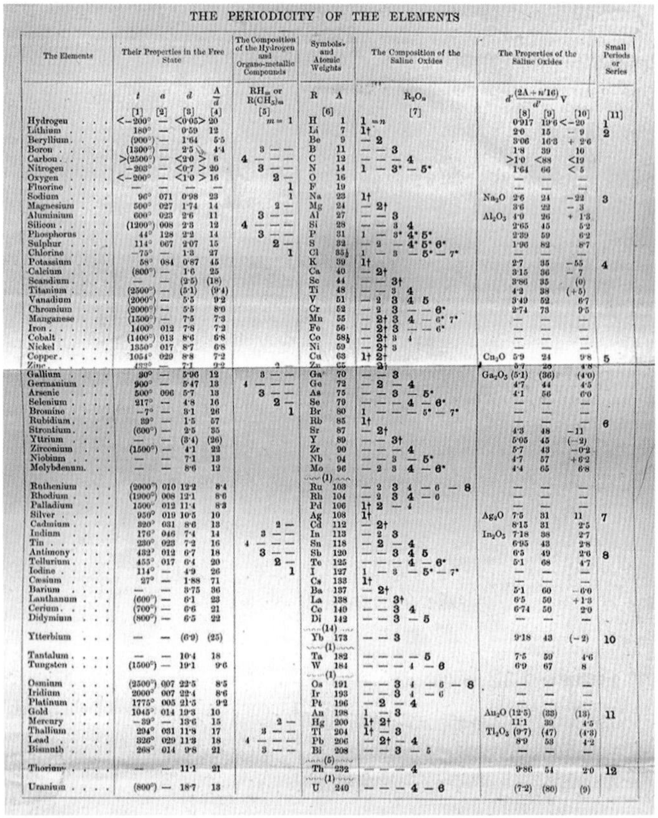

The way that the elements behave – their **properties** – depends on the parts in each atom. For example, their **mass** (or 'atomic weight') depends on how many nucleons they have (because electrons are almost without mass, as they are very, very small compared to nucleons). However, the way the atoms **interact** with other atoms depends on the number and configuration of their electrons.

The first person to notice that the elements could be 'grouped' according to their properties was the Russian, Dmitri Mendeleev. It is said that Mendeleev had a dream in which he 'saw' the pattern that became the **Periodic Table**! Mendeleev noticed a pattern in the way the elements reacted with other elements. In 1869, he published his 'periodic table' and predicted the properties of an element that seemed to be missing from the pattern. In 1875, the element was discovered and named Gallium: its properties turned out to be just as Mendeleev had predicted.

■ **Figure 1.17** An 1891 English version of Mendeleev's first Periodic Table

The Periodic Table

1 H																	2 He
3 Li	4 Be											5 B	6 C	7 N	8 O	9 F	10 Ne
11 Na	12 Mg											13 Al	14 Si	15 P	16 S	17 Cl	18 Ar
19 K	20 Ca	21 Sc	22 Ti	23 V	24 Cr	25 Mn	26 Fe	27 Co	28 Ni	29 Cu	30 Zn	31 Ga	32 Ge	33 As	34 Se	35 Br	36 Kr
37 Rb	38 Sr	39 Y	40 Zr	41 Nb	42 Mo	43 Tc	44 Ru	45 Rh	46 Pd	47 Ag	48 Cd	49 In	50 Sn	51 Sb	52 Te	53 I	54 Xe
55 Cs	56 Ba	57–71	72 Hf	73 Ta	74 W	75 Re	76 Os	77 Ir	78 Pt	79 Au	80 Hg	81 Tl	82 Pb	83 Bi	84 Po	85 At	86 Rn
87 Fr	88 Ra	89–103	104 Rf	105 Db	106 Sg	107 Bh	108 Hs	109 Mt	110 Ds	111 Rg	112 Cn	113 Uut	114 Fl	115 Uup	116 Lv	117 UUs	118 Uuo

57 La	58 Ce	59 Pr	60 Nd	61 Pm	62 Sm	63 Eu	64 Gd	65 Tb	66 Dy	67 Ho	68 Er	69 Tm	70 Yb	71 Lu
89 Ac	90 Th	91 Pa	92 U	93 Np	94 Pu	95 Am	96 Cm	97 Bk	98 Cf	99 Es	100 Fm	101 Md	102 No	103 Lr

■ **Figure 1.18** A modern version of the Periodic Table. Chemical elements are indicated by their letter symbols, and numbers give the number of protons in each element's nucleus

EXTENSION

Explore further! For more information about the modern periodic table of the elements, try these interactive versions online:

- **http://www.ptable.com/**
- **http://www.infoplease.com/periodictable.php**

What is the relationship between atoms and the things atoms make?

So, how does the smallest stuff affect the largest? And what relationship is there between atoms and the larger things that are made from them?

Let's take one atom as an example. **Carbon** is a very important element to us. The great majority of our body is built from molecules that contain carbon. A **molecule** is the smallest particle of a substance that retains all the properties of the substance and is composed of one or more atoms.

Carbon has some interesting properties. Carbon atoms can arrange themselves in different ways to make different molecules with very different properties.

If carbon atoms arrange themselves in a 'tetrahedral' shape, like in Figure 1.19, then the material looks like Figure 1.20.

In this arrangement carbon forms **diamond** – one of the hardest materials known to humanity; it's also transparent and reflective (shiny).

On the other hand, carbon can also arrange itself in flat layers, as in Figure 1.21, and the material then looks like Figure 1.22.

This material is called **graphite**. It is very soft, an opaque (solid) black colour, and it conducts electricity somewhat.

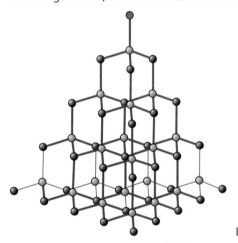

■ **Figure 1.19**

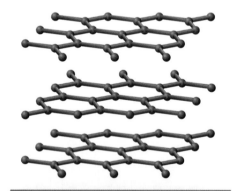

■ **Figure 1.21**

■ **Figure 1.20**
Diamond 'in the rough'

■ **Figure 1.22**
Carbon in the form of graphite

ACTIVITY: Physical properties investigation

Look again at the molecular structures of diamond and of graphite. What is different about their structures?

Figure 1.23 shows some two-dimensional (flat) structural shapes found in molecules.

■ **Figure 1.23**

We have seen that simply changing the arrangement of the atoms that constitute a molecule can completely change the properties of a material.

What properties of materials are there? Look at the properties in the box below. If you are unsure what any of these mean, **find out**.

Some physical properties of materials		
hardness	melting point	viscosity
elasticity	freezing point	plasticity

We can visualize the process of an experiment as a **cycle** because, at the end, we return to the question better informed, and so ask better questions next time. Many of our **Approaches to Learning** can be visualized in this way. Use the **experiment investigation cycle** (Figure 1.24) to help you design and carry out an investigation into the properties of materials.

Safety: Make sure you check your design with your teacher for safety before starting.

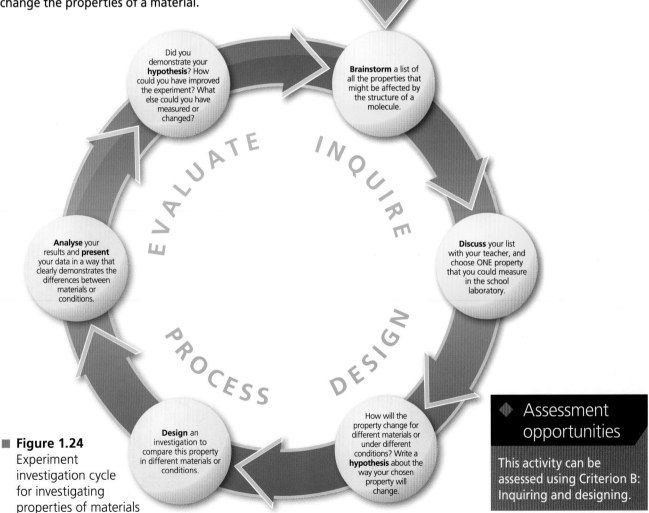

■ **Figure 1.24** Experiment investigation cycle for investigating properties of materials

Did you demonstrate your **hypothesis**? How could you have improved the experiment? What else could you have measured or changed?

Brainstorm a list of all the properties that might be affected by the structure of a molecule.

Discuss your list with your teacher, and choose ONE property that you could measure in the school laboratory.

Analyse your results and **present** your data in a way that clearly demonstrates the differences between materials or conditions.

How will the property change for different materials or under different conditions? Write a **hypothesis** about the way your chosen property will change.

Design an investigation to compare this property in different materials or conditions.

EVALUATE   INQUIRE   PROCESS   DESIGN

◆ Assessment opportunities

This activity can be assessed using Criterion B: Inquiring and designing.

In 2003, a physics professor at the University of Manchester in the United Kingdom gave one of his students a challenge: to make the thinnest piece of graphite he could. The student did quite well, using normal tools to slice the graphite thinner and thinner, and he made a piece of graphite that was between 100 and 1000 carbon atoms thick.

But the physics professor had a better idea – and it was very simple. Instead of cutting the carbon, he used sticky tape to peel away layers of graphite. With a lot of care – and a certain amount of scientific research – Professor Andre Geim managed to produce layers that were less than 10 atoms thick!

These layers formed an amazing new material, now called graphene. Graphene is the thinnest and lightest material ever made – yet it is 300 times stronger than steel and harder than diamond! Graphene is transparent, bendable, and conducts electricity – so one of its first uses will be for touch-screens on portable devices like mobile 'phones or tablets.

For their amazing discovery, Professor Geim and his colleague Konstantin Novoselov were awarded the 2010 Nobel Prize for Physics.

■ **Figure 1.25** An experimental 'flexible' computer screen made from a sheet of graphene

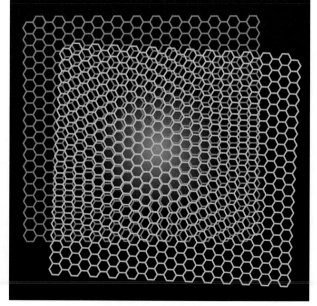

■ **Figure 1.26** The discoverers of graphene, Geim and Novoselov, and its structure

The uses of graphene are still being worked out, so watch out for it in the news. To find out more, try these links:

- www.graphene.manchester.ac.uk/
- http://gigaom.com/2013/07/15/what-is-graphene-heres-what-you-need-to-know-about-a-material-that-could-be-the-next-silicon/

And for an interesting short video: http://youtu.be/dTSnnIITsVg

ACTIVITY: Graphene, super-material for the future?

ATL

- **Information literacy skills**: Access information to be informed and inform others

Take action

! **Study graphene**: carbon super-material for the future!

Can you think of an application of graphene that might be able to help those who live in the less economically developed parts of the world?

Design an information leaflet or an internet 'infomercial' to campaign for investment in the use of graphene to help others.

In your leaflet or infomercial:

- **Describe** and **explain** the science behind the development of graphene as a material.
- **Discuss** and **evaluate** the implications of graphene for commercial use, and consider what other effects graphene might have on our world.
- Be sure to use scientific terminology accurately and document all your sources using a recognized referencing and citation standard.

Assessment opportunities

This activity can be assessed using Criterion D: Reflecting on the impacts of science.

SOME SUMMATIVE PROBLEMS TO TRY

Use these problems to apply and extend your learning in this chapter. The problems are designed so that you can evaluate your learning at different levels of achievement in Criterion A: Knowledge and understanding.

THIS PROBLEM CAN BE USED TO EVALUATE YOUR LEARNING IN CRITERION A TO LEVEL 1–2

1 a Interpret Robert Hooke's diagram and the scale given to **estimate** the length of the flea's hind leg.

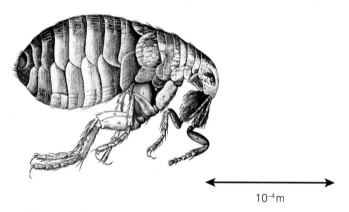

10^{-4} m

■ **Figure 1.27**

b If your eye has a resolution of 0.5×10^{-3} m, **calculate** the magnification of lens needed to be able to clearly see the flea's leg.

c State a suitable instrument to use for this.

THIS PROBLEM CAN BE USED TO EVALUATE YOUR LEARNING IN CRITERION A TO LEVEL 5–6

2 Describe how you would make observations of each of the objects below.

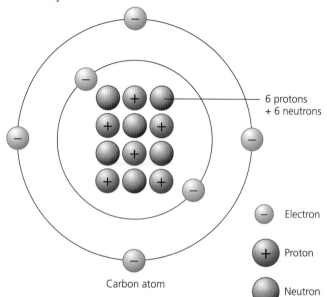

6 protons
+ 6 neutrons

⊖ Electron

⊕ Proton

◯ Neutron

Carbon atom

■ **Figure 1.28**

Using *Powers of Ten* or one of the other websites from earlier in this chapter, **estimate** the powers of ten for measurements in the table below (the first one is done for you).

Object	Diameter (m)
nucleus	10^{-15}
atom	
Sun	
Solar System	
edge of Milky Way galaxy	

Estimate how many atoms can fit, end to end, in a piece of material 1 mm long.

Estimate the volume of a sphere of the same material, and so deduce how many atoms would fit inside the sphere.

A scientist makes the following proposal: 'The scale of things in the Universe follows a regular pattern. For example, the ratio of the size of an atom compared to the size of its nucleus is about the same order of magnitude as the ratio of the size of the Solar System to the size of the Sun.' **Analyse** and **interpret** the data in the table and so **state** whether you think the scientist is correct, or not.

THIS PROBLEM CAN BE USED TO EVALUATE YOUR LEARNING IN CRITERION A TO LEVEL 7–8

3 The diagrams in Figure 1.29 all show arrangements of carbon atoms.

   a **Analyse** the diagrams to **suggest** patterns that they have in common, and any differences you can see between the different arrangements.

      The first diagram shows the structure of diamond. The last diagram shows the structure of graphite. **Compare** the following physical properties of diamond and of graphite: hardness, malleability (flexibility), strength.

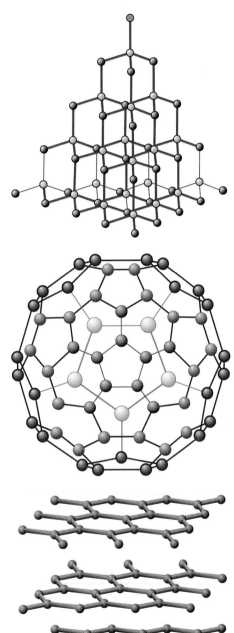

■ **Figure 1.29**

b **Explain** the differences in the physical properties of diamond and that of graphite in terms of the arrangements of atoms shown.

The second diagram shows a molecule known as buckminster-fullerene (Carbon-60).

Suggest what properties you think this substance will have, in comparison to diamond and graphite. **Explain** your reasoning.

Graphene is a form of carbon that consists of very thin layers of only one atom thickness.

Estimate the average thickness of a layer of graphene.

c Artists' pencils often use graphite. If a pencil line is 10^{-4} m thick, **calculate** how many layers of graphite there are on the paper.

d Both graphite and graphene are semiconductors, which means they can conduct electricity, although their resistance is relatively high when compared to conductors like metals. Diamond is not a conductor of electricity (it is an insulator). Electrical conduction occurs where movable electrons are available to carry electric charge.

With reference to the differences in the molecular structures given above, **explain** the differences in conductivity of these different forms of carbon.

Reflection

In this chapter we have explored ways in which we have extended the range of observations that we can make using instruments such as microscopes and telescopes. We have seen how new observations at different scales have revealed patterns and structures in the way forces and matter work together at all scales. We have discussed how humanity elaborates special techniques for isolating variables with a scientific method, and we have used those techniques to relate the properties of materials to their underlying structure.

Use this table to reflect on your own learning in this chapter		
Questions we asked	Answers we found	Any further questions now?
Factual: What is the smallest thing? What is the largest thing? How do we measure them? What is the Universe made from?		
Conceptual: How is the Universe structured? How are the smallest and the largest things in the Universe connected?		
Debatable: What would the world look like if you were very small? Or very large?		

Approaches to learning you used in this chapter	Description – what new skills did you learn?	How well did you master the skills?			
		Novice	Learner	Practitioner	Expert
Information literacy skills					
Critical-thinking skills					
Transfer skills					
Communication skills					

Learner profile attribute(s)	How did you demonstrate your skills as an inquirer in this chapter?
Inquirers	

2 How do forces and matter interact?

Through identifying **relationships** of **similarity and difference**, we understand how force and matter **interact**.

■ **Figure 2.1** A touch brings electrical forces to interact

CONSIDER AND ANSWER THESE QUESTIONS:

Factual: What forces are there? How strong are different forces? What is the strongest force? What is the weakest force?

Conceptual: How do forces and matter interact? How far do forces reach? Where do forces begin and where do they end? What holds the Universe together? What stops the Universe from collapsing in on itself?

Debatable: Is any force more important than others across the Universe? Can we control natural forces?

Now **share and compare** your thoughts and ideas with your partner, or with the whole class.

IN THIS CHAPTER, WE WILL ...

■ **Find out** how observations of nature have led us to identify fundamental forces that interact with matter in different ways.

■ **Explore** three kinds of force closely: gravitation, electrical and magnetic forces, and analyse how these forces affect the matter around them.

■ **Take action** and explore the way metals can be recycled using electromagnetic force.

These Approaches to Learning skills (ATL) will be useful ...

■ Critical-thinking skills

■ Information literacy skills

■ **Figure 2.2** Machines designed to utilize interacting forces

In each of the examples in Figure 2.2, a different force is being used to do useful work. Can you identify the forces in action?

What makes everything 'stick' together? This might seem like a strange question, but in the last chapter we looked at how humanity's ideas about matter – the 'stuff' in the Universe – have changed, but have settled on a picture of matter that is made up of smaller and smaller pieces.

If we think about atoms for a moment, what stops atoms coming apart, like a pile of flour? Why, instead, do they stick together, like baking dough?

On the other hand, what stops **everything** from sticking together? If something makes atoms stick together, why don't **all** atoms stick together, so that the entire Universe becomes one big lump of stuff with nothing on the outside?

For many years the answer to these questions was 'because some matter wants to stick together, and some matter doesn't!' Philosophers, such as the ancient Stoics of Milesia in Greece, believed that matter was 'alive' and had its own 'spirit' or 'will'. This idea persisted for a long time in European culture, passed down particularly in the works of Aristotle. The idea of matter having a 'will' or a 'mind' of its own was also used to explain the way things moved or slowed down – as we will see in the next chapter.

Arguably, it was the great mathematician and philosopher Galileo Galilei (1564–1642) of Pisa, in what is now Italy, who suggested that we ought to think of the Universe differently: not in terms of 'matter with a mind of its own', rather in terms of matter as inanimate (unmoving) 'stuff' with separate 'forces' that act on it.

All forces can be attractive or repulsive (Figure 2.3).

But not all forces show both of these properties to the same degree!

● **We will reflect on this learner profile attribute …**

● Thinker – we will be using critical-thinking skills to analyse observations and deduce properties of force fields.

◆ **Assessment opportunities in this chapter**

◆ **Criterion A**: Knowing and understanding
◆ **Criterion C**: Processing and evaluating
◆ **Criterion D**: Reflecting on the impacts of science

KEY WORDS

attract	field	repel

Attractive –
the force tries to bring objects together

Repulsive –
the force tries to push objects further apart

■ **Figure 2.3**

What forces are there? How do they interact with matter?

FORCES IN ACTION

Forces act around the objects that cause them.
Each of the different forces originates in a different kind of object.

■ **Table 2.1** Forces and their origins

Force	Origin (what causes it)	Nature	Properties
electromagnetic – electrical	electrons protons	electrons carry 'negative' charge protons carry 'positive' charge	electric charges interact in different ways: same charges are repulsive: + and +, or – and – different charges are attractive: + and –
electromagnetic – magnetic	electrons in groups of atoms	groups of atoms form magnetic 'dipoles' with a 'north' end (or pole) and a 'south' pole	magnetic dipoles interact in different ways: same poles are repulsive: N and N, or S and S different poles are attractive: N and S
gravity	all mass	matter attracts matter	only attractive gravitational forces have so far been observed

ACTIVITY: A forces smorgasbord

Try these simple experiments to investigate forces in everyday action.

ATL

■ **Critical-thinking skills:** Evaluate evidence and arguments

1 a Take a balloon and blow it up. Tear up some tissue paper into small pieces, about 0.5 cm square. Rub the balloon on the sleeve of a jersey or coat for about a minute.
  b Predict what will happen when you bring the balloon near to, but not touching, the tissue paper.
  c Observe what happens to the tissue paper.
2 a Now take a plastic drinking straw and a small (250 ml) water bottle (a full one is better, since it won't fall over). Balance the drinking straw on top of the water bottle. Take a plastic ruler and rub it on the sleeve of a jersey or coat for about a minute.
  b Predict what will happen when you bring one end of the plastic ruler near to, but not touching, the drinking straw.
  c Observe what happens to the drinking straw.
3 a Take some grapes, and a strong magnet. (If you don't have a 'laboratory' style magnet, you may find that strong refrigerator magnets will work, especially the ones in the shapes of letters or numbers!)
  b Predict what will happen when you place a grape or two on the table and move the magnet over the top.
  c Suspend one of the grapes from a lab stand using a thin piece of thread and a little tape or glue.
  d Ensure the grape is completely still and not twisting on the thread, then move the strong magnet close to it. Observe what happens to the grape.

DISCUSS

Compare the different properties of the three example forces (gravitational, electrical, magnetic) using the information in Table 2.1. Organize the properties using a Venn diagram. Place forces that attract only in one circle, forces that repel only in the other circle, and any forces that both attract and repel in the intersection.

Which force is the 'odd one out'?

> Identifying similarities and differences between things is a good way to see patterns and connections – and these help us to understand the bigger picture.

■ **Figure 2.4** Organizing forces visually

Physicists have identified **four** fundamental (basic) forces at work in the Universe. In the above experiments, you have investigated the effects of **two** of them: **gravitational** force (gravity), and **electromagnetic** force. However, electromagnetic force can be 'split' and observed as separate **electrical** and **magnetic** effects.

The other two forces are a little more difficult for us to measure with simple experiments: they are the **strong nuclear force** and the **electroweak force**. These forces only affect the nuclei of atoms and sub-atomic particles. We will return to these in Chapter 11.

4   a   Take a lemon, and a small coin. Place the lemon on its side and try to balance the coin on top of it. Easy, isn't it? Now put the lemon in a bowl of water.
    b   Predict what will happen when you try to balance the coin on top of the lemon now.
    c   Observe what happens to the coin.
5   Now **organize** your observations using the following design for a table. Try to suggest 'provisional' explanations for what you have observed. The first one has been done for you.

Inquiry in science

Science is all about inquiry – but how do we know what questions to ask? Scientific inquiry often starts with some **preliminary observations** which lead us to make a **hypothesis** about how things work. We can then design an investigation to test the predictions we make.

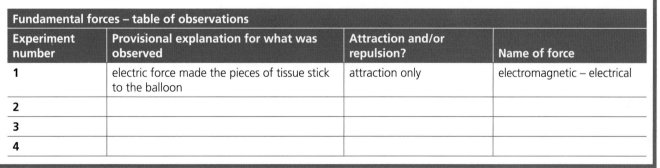

Fundamental forces – table of observations			
Experiment number	Provisional explanation for what was observed	Attraction and/or repulsion?	Name of force
1	electric force made the pieces of tissue stick to the balloon	attraction only	electromagnetic – electrical
2			
3			
4			

How far do forces go? What is the strongest force? What is the weakest force?

We can only observe the effects of forces indirectly.

The easiest force fields to observe are magnetic fields. The element iron has a strong tendency to form magnetic dipoles and this means that iron exhibits strong magnetism in nature – this is called ferromagnetism.

The Earth itself has a magnetic field due to the presence of iron. The Earth's magnetic north pole is the region towards which all compasses point. Interestingly, this is not in the same place as the geographic North Pole, which is where the Earth's axis of rotation passes through the planet's surface (see Figure 2.6).

The existence of natural magnets has probably been known by humanity for thousands of years. In the Islamic tradition, the North Pole is often associated with Jabal Qaf, or the furthest point of the Earth – Arab sailors are known to have used primitive compasses to navigate.

■ **Figure 2.5** Haematite is an ore of iron that is non-magnetic. Other ores, such as Magnetite, exhibit 'natural' magnetism.

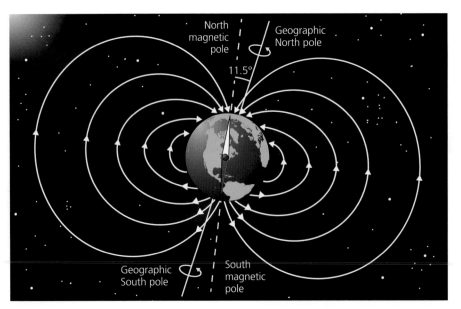

If one end of a compass needle points towards the Earth's magnetic north pole, which pole is located at that end of the needle?

■ **Figure 2.6** The Earth has its own magnetic field, which behaves like a large dipole or bar magnet through the centre of the planet

SEE-THINK-WONDER

Look at Figures 2.7 and 2.8.

What do you see? What does this image make you think? What do you wonder about it?

■ **Figure 2.7** Compasses like this one have been identified as in China from as far back as 220 BCE

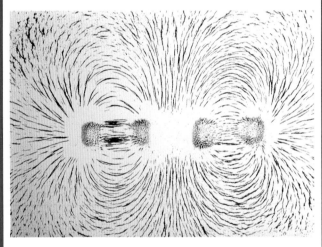

■ **Figure 2.8** A bar magnet will cause iron filings to orientate along lines of equal magnetic strength

ACTIVITY: Observing the form of magnetic fields

■ ATL

- **Critical-thinking skills**: Evaluate evidence and arguments

1 Place a magnet under a piece of thin card, and then use a pepper-pot or shaker to 'sprinkle' magnetic filings over the region where the magnet is concealed.
2 Observe the patterns produced in the filings in these different cases:
   a a single bar magnet or dipole
   b two same poles close together
   c two different poles close together.
3 **Summarize** your observations. How does the field differ for different kinds of interaction? Use the words below and a table like the one shown in your summaries.

attractive	repulsive	approach	avoid

For a single magnet	
For two same poles	
For two different poles	

Since iron is so strongly magnetic, small pieces of it will 'align' themselves with the lines of force in a nearby magnetic field. We can use this fact to observe the form of magnetic fields.

We can represent a magnetic field by using **force arrows** which point from North to South (Figure 2.9).

■ **Figure 2.10** A ferrofluid is a suspension of iron particles in oil, giving a 'magnetic' fluid

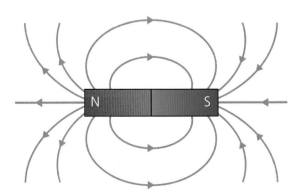

■ **Figure 2.9** Representing magnetic fields using force arrows

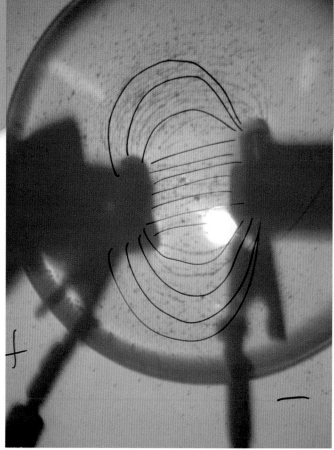

■ **Figures 2.11** and **2.12** In these photos two copper plate 'electrodes' are being submersed in vegetable oil. The electrodes are then connected to a high-voltage supply of electricity, and black pepper is sprinkled into the oil. The pepper aligns itself with the electric field in the oil

It is rather more difficult to observe electrical fields in action since this requires quite large quantities of electrical charge. The photographs (Figures 2.11 and 2.12) show an experiment using high voltages to observe the effect of an electric field between two copper plates. Your teacher may be able to show you this in practice.

Electric fields are said to begin on regions of positive charge and end on regions of negative charge.

When the charges are not moving, the field is said to be **static**.

DISCUSS

Figure 2.13 shows the field around two different charges.

Compare the diagram for two different electric charges to your observations for magnetic dipoles in the previous activity on observing the form of magnetic fields.

Predict the appearance of the field for two same charges.

Explain why you think this.

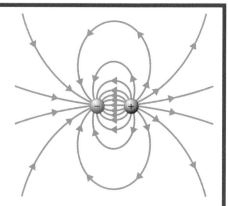

■ **Figure 2.13** Static electric field around attracting charges

Sometimes the charges can move around within a particular material, without leaving the material itself. For example, consider our earlier experiment with the straw and bottle (Figure 2.14).

Gravity is rather the 'odd one out.' Certainly, gravitational force travels through space and creates a force field. However, as far as we know there is only one gravitational 'pole' or ' charge' – all mass behaves in the same way gravitationally, and only attracts other mass.

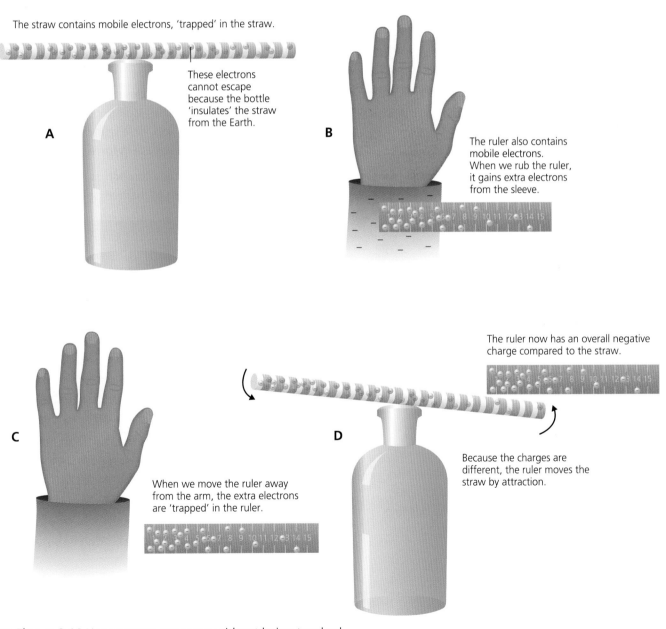

The straw contains mobile electrons, 'trapped' in the straw.

These electrons cannot escape because the bottle 'insulates' the straw from the Earth.

A

B

The ruler also contains mobile electrons. When we rub the ruler, it gains extra electrons from the sleeve.

The ruler now has an overall negative charge compared to the straw.

C

D

When we move the ruler away from the arm, the extra electrons are 'trapped' in the ruler.

Because the charges are different, the ruler moves the straw by attraction.

■ **Figure 2.14** How a straw can move without being touched

ACTIVITY: Measuring Earth's gravitational field strength

A force or 'Newton' meter – sometimes called a spring balance – uses a spring adjusted to stretch a known amount for a given gravitational force.

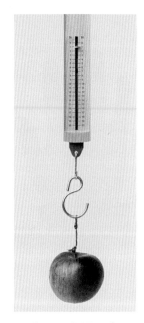

Method

You will need:
- **Masses of known quantity (for example, 10 g, 50 g, 100 g)**
- **A force meter**

1 Hang the force meter vertically from a lab stand.
2 Add masses to the force meter at regular intervals.
3 Measure the force in Newtons produced by each mass you add to the meter.

■ **Figure 2.15** A force or 'Newton' meter

Variables

In your experiment plan, clearly identify the variables that are:
- **controlled** (those changed, and those kept the same)
- **measured**.

Results

Record your results clearly in a table like this one, showing the units of measurement in the heading.

Mass *m* (g)	Force *F* (N)

Analysis

Show your results on a graph with the **independent variable** on the *x*-axis and the **dependent variable** on the *y*-axis. (Unsure which variable is which? See Chapter 1!)

The points on your graph should be in more or less a straight line, although the line might 'wiggle' somewhat due to **scatter** in the points.

What might have caused the scatter in the measurements?

Since the points appear to be so close to a straight line, we can probably make the assumption that the relationship between mass and force is **linear** – that is, **each mass gives the same increase in force**.

So, we can draw a 'best-fit' line through as many of the points as possible, or as close to them as possible:

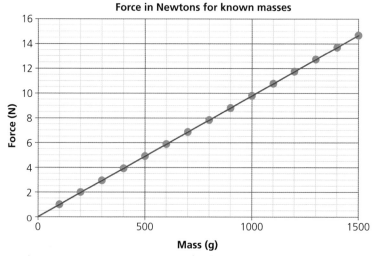

■ **Figure 2.16** Force in Newtons for known masses

Using this straight line, we can now figure out the **relationship** between the mass and the force – that is, how much the force increases for each additional mass. We can find this by finding the gradient or slope of the line on the graph:

$$\text{Gradient} \frac{\Delta y}{\Delta x} = \frac{\text{change in } y}{\text{change in } x} = \frac{\text{change in force}}{\text{change in mass}}$$

▼ Link: Mathematics

The Greek symbol Δ or 'delta' is often used in equations to represent the **change** in a variable. For example,

$$\Delta x = x_2 - x_1$$

Where x_2 and x_1 are different values of the variable x.

One way to find the gradient of a line is to draw a right-angled triangle on the graph, as large as possible, so that it encompasses the greatest possible range of results (Figure 2.17).

Use your graph to **calculate** the gradient of your line.

Conclusion

State the relationship between **mass** in grammes and **gravitational force**.

Evaluation

Evaluate your results. How sure are you that your experiment data gave a reliable result? (Clue: what about that 'wiggle' and 'scatter'?) How could you improve the reliability of the data?

Remember that an **evaluation** is a specific thing in science. It doesn't mean 'work out a value' (as in maths), nor does it mean 'reflect on your progress'. In science, we are evaluating how well the results of our experiment (our observations) actually support the conclusions we draw from them.

◈ Assessment opportunities

In this activity you have practised skills that are assessed using Criterion C: Processing and evaluating.

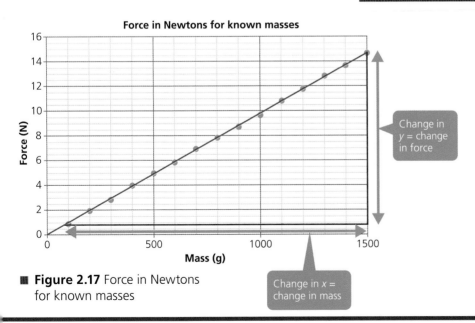

■ Figure 2.17 Force in Newtons for known masses

where

$g = 9.81$ Newtons per kilogramme, $N\,kg^{-1}$, on Earth

Notice that the gravitational field strength g for the Earth is actually a little less than 10 Newtons for every kilogramme of mass.

Of course, we are used to thinking of our 'weight' in kilogrammes – however, this is a bit misleading. To a physicist, kilogrammes are a measure of **mass**, the amount of material in your body, and not of the gravitational force produced on it.

The Newton gravitational force is referred to as the weight of an object.

weight = mass × gravitational field strength

w (N) = m (kg) × g

ACTIVITY: Analysing gravitational fields

■ ATL

■ **Critical-thinking skills**: Interpret data; Draw reasonable conclusions and generalizations

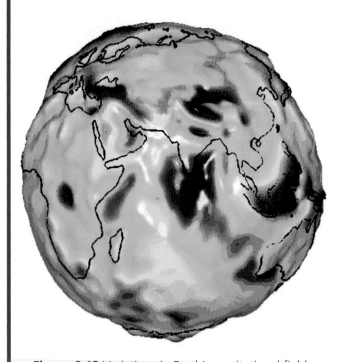

■ **Figure 2.18** Variations in Earth's gravitational field measured by NASA's GRACE satellite

Look at the image from the GRACE satellite. The colours on the image show variations in gravitational field. Red regions indicate relatively high gravitational fields, yellow a moderate field, and dark blue lower gravitation.

1 **Suggest** how your weight would be different in the blue and the red regions.
2 **State** what effect the different fields would have on your mass.
3 **Suggest** what you think might be causing the different regions of gravitational field strength.

Using the equation for gravitational force, **calculate** your weight if you were standing on the surface of some different objects in our Solar System. Gravitational field strength g on each object is given in the table.

Place	Gravitational field strength g at surface ($N\,kg^{-1}$)
Moon	1.62
Mars	3.73
Jupiter	25.94
Asteroid Ceres	0.26

◆ Assessment opportunities

In this activity you have practised skills that are assessed using Criterion A: Knowledge and understanding.

How do forces and matter interact?
How far do forces reach?

33

ACTIVITY: Lift off! Analysing the variation of gravitational field strength with distance

ATL

- **Critical-thinking skills:** Use models and simulations to explore complex systems and issues; Identify trends and forecast possibilities; Interpret data, draw reasonable conclusions and generalizations

When spacecraft leave the Earth's surface, they have to work against the Earth's gravitational attraction. Fortunately, the further they climb away from the Earth's mass, the weaker that gravitational field strength becomes.

Measuring field strength is tricky. One way to do this is to use the variation in **weight** of a known mass at different distances above the Earth's surface.

The graph in Figure 2.20 shows the measurements made by a deep-space probe as it travels away from the Earth. On board the probe were the following apparatus:

- a sensitive balance that continually measures the weight of a 1 kg mass
- a radar altimeter, a device that measures distance from the Earth's surface using radio waves
- a radio transmitter to send the measurements back to the experimenters.

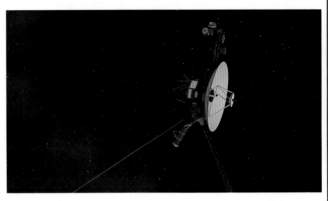

Figure 2.19 The Voyager space probes, launched in the 1970s, have now travelled to the limits of our Solar System

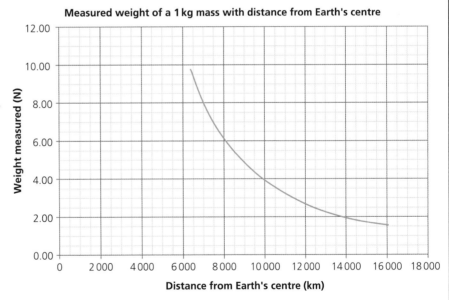

Figure 2.20 Measured weight of a 1 kg mass with distance from Earth's centre

Imagine you are the space scientist whose job it is to analyse this data. Use the headings and guiding questions below to help you write a report.

Data analysis

Interpret and **analyse** the graph to find:

- **the change in weight between 6400 km and 8000 km**
- **the change in weight between 14000 km and 16000 km.**

Conclusion

Write a conclusion in which you describe the way in which the weight changes as the distance increases.

Evaluation

Explain why the graph doesn't start at a distance of 0 km.

As the probe travels into deep space, it may be affected by the gravitational fields of other objects, such as the Moon or the planet Mars. **Outline** how this might affect the data.

Evaluate the experiment – what problems might there be with the data? Are the data reliable? Are the data valid?

Suggest what changes could be made to the experiment to eliminate the problems you have identified above.

◆ Assessment opportunities

This activity helps you practise skills that are assessed using Criterion C: Processing and evaluating. (However, for full assessment in Criterion C you would have to also collect and present the data yourself – which might be tricky if you don't have a space probe and rocket to hand….)

The strength of gravitational and electrical fields varies as the 'reciprocal' of square of the distance, or

$$\text{field strength} \propto \text{(is proportional to) } \frac{1}{d^2}$$

So, if the field strength is 1 unit for a distance of d m,

and the distance doubles to $2d$, the field strength $= \frac{1}{(2)^2}$ what it was before

so field $= \frac{1}{4}$ original strength.

If the distance triples to $3d$, then the field strength $= \frac{1}{(3)^2}$ what it was before,

so field $= \frac{1}{9}$ strength

… and so on.

This is called an **inverse square relationship**. This relationship for field strength and distance from the origin of the field holds true for other fields – electrical and gravitational – also.

But how strong are the forces in the first place?

In fact, the fundamental forces have very different strengths relative to each other (Table 2.2)

■ **Table 2.2** Relative strengths and distances of action for fundamental forces

Force	Relative strength (compared to strong nuclear = 1)	Distance of action (m)
strong nuclear	1	$<10^{-15}$ = less than size of a nucleus
electromagnetic	$1/137 = 7.2 \times 10^{-3}$	infinite
weak	approximately 10^{-6}	10^{-18} = approximate size of a proton
gravity	6×10^{-39}	infinite

THINK-PAIR-SHARE

Using the data in Table 2.2, and the equations for distance above, **discuss** these questions with your partner and feed back to class:

- **Why don't we feel each other's force of gravitational attraction pulling us together?**
- **How can stars like the Sun attract large objects like planets?**
- **When we jump off a stool, what force field is accelerating us to the Earth?**
- **When we land on the ground, what force field is stopping us?**
- **Which force works in the shortest distance? Why?**

Where do forces begin? Where do they end?

At very large distances, the force from an electric charge, a mass or a magnet becomes too weak for us to measure it with our simple apparatus. Does this mean that the force field has disappeared?

What happens to the force as the distance gets larger?

In fact, there is no real value of distance d for which our force field equation

$$\text{field strength} \propto \frac{1}{d^2}$$

gives the result field strength = zero.

So, do forces ever end?

ACTIVITY: Modelling force fields

■ ATL

- **Information literacy skills**: Organize and analyse data using digital tools

What happens to force fields as the distances from the origin get very large?

Use a spreadsheet or graphic display calculator to work out the size of the field strength for distances that are larger and larger. (Your teacher will be able to help you with this if you are unsure of what to do.)

▼ Link: Mathematics

Find out about **asymptotes** in mathematics. What other functions are **asymptotic**?

ACTIVITY: Recycling – a win-win for the planet

Metals are expensive to extract from their ores, and it is increasingly common for metal in domestic and industrial waste to be recycled. Ferromagnetic metals are usually separated from the rest of the waste using their magnetic properties. The waste is passed along a conveyor belt underneath another belt which contains strong electromagnets that attract the ferrous waste to them.

Are there metal-recycling facilities in your area? If not, why not?

Find out about metal recycling. Use some of these to start your research:

metal prices

metal recycling prices

metal recycling impacts

metal mining impacts

Here are some ideas on how you can **take action** to help conserve the Earth's metal resources:

- **Produce an information leaflet, web page or poster to raise awareness about metal recycling. Research the way that one metal is mined and extracted, and how the science and technology of metal recycling might help reduce the impact of this process.**
- **Write your research findings in the form of a report to your government, and, at the end, make a recommendation on this question: should the government invest in more metal recycling?**
- **Find out where metals are recycled near you. Start a metal can recycling programme in your school or local community.**

■ **Figure 2.21** Metal recycling

SOME SUMMATIVE PROBLEMS TO TRY

Use these problems to apply and extend your learning in this chapter. The problems are designed so that you can evaluate your learning at different levels of achievement in Criterion A: Knowledge and understanding.

THIS PROBLEM CAN BE USED TO EVALUATE YOUR LEARNING IN CRITERION A TO LEVEL 3–4

1 Figure 2.22 shows how the gravitational force (weight) varies with mass on the planet Mars.

  a **Outline** the difference between the physics concepts of mass and weight.

  b Using the data earlier in the chapter, **sketch** similar lines on the graph for the Moon, Jupiter and for the Asteroid Ceres.

  c If an astronaut weighs 700 N on Earth, **calculate** what they would weigh on the Moon.

THIS PROBLEM CAN BE USED TO EVALUATE YOUR LEARNING IN CRITERION A TO LEVEL 5–6

2 Figure 2.23 shows four lines which might represent the way the strength of a force field changes with distance.

  a **State** which line (A, B, C, D) on the graph most closely shows the way in which an electric field varies with distance.

  b **Describe** the variation.

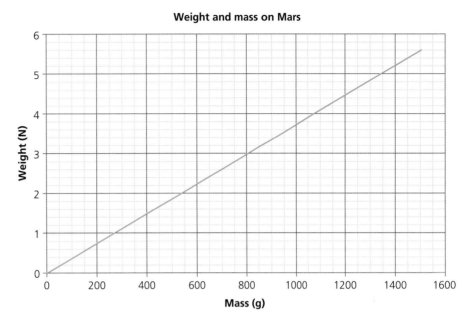

Figure 2.22 Weight and mass on Mars

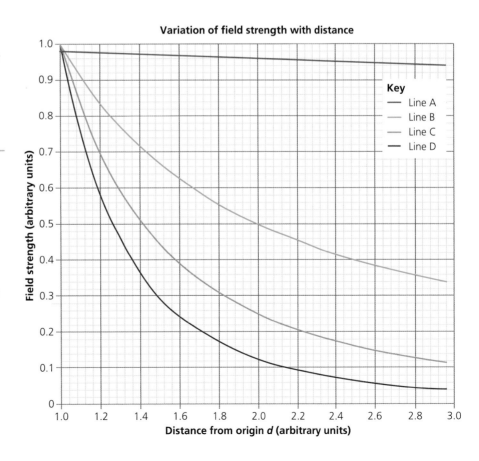

■ **Figure 2.23** Variation of field strength with distance

3 MagLev trains work by lifting a train on electromagnetic fields to hover a short distance above a track. The magnets in the train and the track are designed to repel each other.

An electromagnet is designed such that the magnetic field produced varies like the electric field in the graph above.

a A MagLev train, when unloaded, hovers a distance *d* above the track. When fully loaded with passengers, its weight is four times greater than the train alone. **Calculate** the new distance of the train above the track in terms of *d*.

b The engineers working on the train notice that as the electromagnets become hotter, they produce a weaker field. **Suggest** which line on the graph might be a suitable model for this field.

c Using the line you have chosen in **b** above, **outline** the effect on the position of the loaded train. Support your answer with **calculations**.

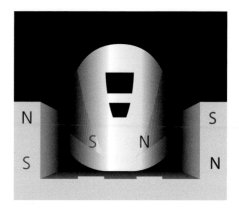

■ Figure 2.24

THIS PROBLEM CAN BE USED TO EVALUATE YOUR LEARNING IN CRITERION A TO LEVEL 7–8

4 Victoria's ambition is to climb Mount Everest, also known in Nepal as Sagarmāthā and in Tibet as Chomolungma. She thinks that the effect of the changing gravitational force will mean that she will find the climb easier as she climbs higher.

The Earth's radius is approximately 6400 km.

The height of mount Everest is approximately 8.8 km.

a **Explain** how the gravitational force between two objects changes with the distance between them.

b With reference to the data above and to what you know about gravitational force fields, **evaluate** Victoria's prediction.

c **Calculate** the height above the Earth's surface at which Victoria's weight would be half its value on the surface.

Reflection

In this chapter we have explored the way forces can cause changes without apparently touching other objects. These are sometimes called non-contact forces, to distinguish them from the effects of forces that make physical contact with objects.

But when you think about it, what is 'contact'? When you touch the table with your finger, the table pushes back, not because the atoms of the table are touching the atoms of your finger, but because the atoms are electrically repelling. In fact, our concept of 'touch' is, perhaps, based on a misunderstanding – we think that matter actually touches matter, when in reality only electric fields interact!

In the next chapter, we will be considering how we can organize contact forces into useful systems and structures.

Use this table to reflect on your own learning in this chapter.					
Questions we asked	Answers we found		Any further questions now?		
Factual: What forces are there? How strong are different forces? What is the strongest force? What is the weakest force? Your questions?					
Conceptual: How do forces and matter interact? How far do forces reach? Where do forces begin, where do they end? What holds the Universe together? What stops the Universe from collapsing in on itself? Your questions?					
Debatable: Is any force more important than others across the Universe? Can we control natural forces? Your questions?					
Approaches to learning you used in this chapter	Description – what new skills did you learn?	How well did you master the skills?			
		Novice	Learner	Practitioner	Expert
Collaboration skills					
Information literacy skills					
Critical-thinking skills					
Reflection skills					
Learner profile attribute(s)	How did you demonstrate your skills as a thinker in this chapter?				
Thinker					

3 Amazing structures: how have we learned to use force?

○ Nature's *forms* have inspired us to use **systems** of force and to create *innovative* structures.

■ **Figure 3.1** A bungee jump

CONSIDER AND ANSWER THESE QUESTIONS:

Factual: What are the effects of forces on the form of objects? What are the strongest shapes?

Conceptual: How do forces work together? How do forces affect form? Why do things balance? What is the relationship between structural form and strength? Why do some buildings stay standing longer than others?

Debatable: Are all the solutions to humanity's problems already to be found in nature?

Now **share and compare** your thoughts and ideas with your partner, or with the whole class.

○ IN THIS CHAPTER, WE WILL …

■ **Find out** what happens when forces are in physical contact with objects and each other.

■ **Explore** the way systems of forces can be controlled using structural forms.

■ **Take action** to find out how international aid organizations use structures to help people who are displaced by disaster.

■ These Approaches to Learning (ATL) skills will be useful...

■ **Creative-thinking skills**

■ **Critical-thinking skills**

■ **Transfer skills**

■ **Information literacy**

● We will reflect on this learner profile attribute …

● Caring – how can the physics of structures be used to help others?

Figure 3.2 A tug of war

Have you ever taken part in a tug of war? Tug of war is an international sport and has its own association. It was even once an Olympic event! The Tug of War Association has a YouTube channel here: **www.youtube.com/user/kcorin1**

Watch one of the videos.

In a tug of war, physical forces are working together. Each team exerts a pulling force on the other, and the object of the game is to pull the opposing team far enough such that a mark on the rope moves past the centre point. In this chapter we will analyse and explore the ways that physical forces can interact.

PULLING IT APART

What is happening in the tug of war when the teams are equally matched, and nobody is moving?

◆ Assessment opportunities in this chapter

◆ **Criterion A**: Analysing force systems, calculating resultant forces

◆ **Criterion B**: Investigating deformation and stretch in a bungee elastic

◆ **Criterion C**: Presenting, interpreting and analysing data, evaluating hypotheses about structures

◆ **Criterion D**: Reflecting on the impacts of science

KEY WORDS

balance	force
equilibrium	resultant

ACTIVITY: Tug-of-war mind experiment

■ ATL

■ **Creative-thinking skills**: Make guesses, ask 'what if' questions and generate testable hypotheses

This is a 'mind experiment', in which we try to figure out what will happen based on our **experience** and on our **imagination**.

With a partner, take a short piece of string and mark its centre point with a felt-tip pen, or similar. Tie the string around your little fingers and pull!

Now **think-pair-share** your ideas about these questions.

When the two teams are each pulling by the same amount and neither is moving,

- **is there a force in the rope?**
- **what will happen if the rope is cut?**
- **if you placed a Newton meter in the rope, what would you predict that it should read?**

How do forces work together?

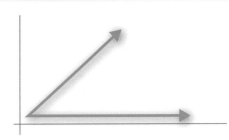

■ **Figure 3.3**

To analyse situations involving forces, we can represent the forces using arrows such that:

■ the arrow points in the direction that the force is acting, starting on the object it is acting on
■ the length of the arrow is proportional to the size of the force.

We might use a scale diagram (Figure 3.3) to make sure our force arrows accurately represent systems of forces – for example, 1 cm = 1 N. We can also define the direction of the forces relative to fixed axes in space – for example the *x*- or *y*-axes.

The following diagrams show the forces in tug-of-war contests. The arrows are drawn to a scale of 1 cm = 20 N. In each case, force towards the **right** is defined as **positive**.

For each diagram, decide:
■ which way the teams will go
■ the size of the force moving them that way (in Newtons).

(The first contest has been done for you!)

Contest 1

$$\text{force } \textbf{left} = 4 \times -20\,\text{N} = -80\,\text{N}$$

$$\text{force } \textbf{right} = 2 \times 20\,\text{N} = +40\,\text{N}$$

$$\text{Total force} = (-80) + 40 = -40\,\text{N}$$

$$\text{direction} = \text{left}$$

Force LEFT = 4 × –20 N
= –80 N

Force RIGHT = 2 × 20 N
= 40 N

Total force = (–80) + 40 = –40 N (direction = left)

■ **Figure 3.4** Tug of war, contest 1

■ **Figure 3.5** Tug of war, contest 2

■ **Figure 3.6** Tug of war, contest 3

▼ Link: Mathematics

■ ATL

■ **Transfer skills:** Make connections between subject groups and disciplines

A force arrow is an example of a **vector**. Vectors are used in mathematics to solve problems in geometry and, more generally, topology (the mathematics of space).

Person A   Person B

This arrow is the force of A pulling on B

This arrow is the force of B pulling on A

Figure 3.7 Force shown in tug of war, contest 1

The blue force arrows in the above diagrams show the force that each person is applying to the rope. However, we might just as well have drawn force arrows to show the force each person experiences due to the other person. Figure 3.7 shows contest 1 drawn this way, instead.

But of course if we work out the sizes and directions of the forces we get:

force **left** = 4 × 20N = −80N

force **right** = 2 × 20N = +40N

Total force = (−80) + 40 = −40N

direction = left

So, it makes no difference to the answer.

The rope can be thought of as experiencing forces in either direction, pulling both ways at once! So, if we placed a Newton meter in the rope here, it would actually show both the forces present:

total force on Newton meter = 80 + 40 = 120N

This force is called the 'strain' in the rope.

In contest 2, the forces from A and B are exactly equal in size, but opposite in direction. In this case, the forces seem to cancel out, and neither team will move. We say that that forces are balanced and that the system is in **static equilibrium**.

Of course, this doesn't mean the forces have disappeared – they still act in the rope, and a Newton meter would measure the strain they produce.

When objects are stationary on Earth, they still have forces acting. We found out in the last chapter that the force of gravity acts everywhere on masses. So any mass within a gravitational field will experience a force due to that field.

For example, we know that any object sitting on a surface experiences the gravitational force acting on its mass as weight, *W*.

So, why don't the books in Figure 3.8 crash through the table? Obviously, because the table stops them by pushing back with an equal and opposite force to the weight. This

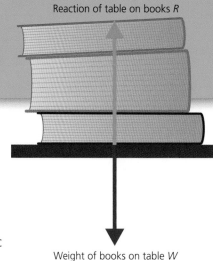

Reaction of table on books *R*

Figure 3.8 Static forces in action

Weight of books on table *W*

'push back' is called the reaction, *R*. If the reaction is in the same axis as the weight, it is also called the normal force, *N*.

You probably noticed that in the tug of war, the forces are always directly opposite to each other – if we choose the *x*-axis to be parallel to the rope, the forces are always parallel to the *x*-axis.

In real life, this isn't always the case …

THINK-PAIR-SHARE

■ ATL

■ **Critical-thinking skills**: Practise observing carefully in order to recognize problems

What would happen to these forces if you leaned down on top of the books?

● **Is it possible for *W* > *R*? What would happen?**
● **Is it possible for *R* > *W*? What would happen?**

Think-pair-share: **Describe** what is happening in these cases (Figures 3.9 and 3.10), in terms of *R* and *W*:
● **'In Figure 3.9, the greatest force is … because …'**
● **'In Figure 3.10, the greatest force is … because …'**

Figure 3.9 Falling through ice

Figure 3.10 Bouncing on a trampoline

How do forces work together?

THE PROBLEM OF THE STUBBORN DONKEY

Andrea has a pet donkey called Tetu. Tetu doesn't like to be taken into his stable to sleep at night and Andrea sometimes has to pull the stubborn mule with ropes.

In fact, sometimes when Tetu really digs his hooves in, Andrea has to call on his father to attach a second rope to the donkey so that they can both pull Tetu into the stable.

The forces on Tetu are shown in Figure 3.12.

How can we work out whether Andrea and his father can overcome Tetu's resistance?

We can use force arrows to solve the problem. If Andrea and his father both pull Tetu in the negative (left) direction; the effect of their forces will combine to give us a **resultant force** in that direction.

The size of the resultant can be found by 'completing the parallelogram' which is made by their two forces (Figure 3.13).

So, the dotted line *a* is the effect of *A* (Andrea's pull) added onto *F* (his father's pull); to look at the same thing another way, the dotted line *f* is the effect of the father's pull added onto Andrea's pull. The lines *f* and *a* do not represent real forces, but the effects of *F* and *A* respectively (Figure 3.14).

The resultant *R* of *F* and *A* is then their combined effect – the black arrow shown (Figure 3.13) which crosses the parallelogram of forces. In vector notation,

$$\vec{R} = \vec{F} + \vec{A}$$

Is the force arrow for *R* greater than *T*? If so, it looks like Tetu is going to get dragged into the stable after all.

You can find some example problems to try at the end of this chapter.

Figure 3.11 Andrea and Tetu

Key
← Father's pull (F)   ← Andrea's pull (A)   → Tetu's resistance (T)

Figure 3.12

Key
← Father's pull (F)   ← Andrea's pull (A)   → Tetu's resistance (T)

Figure 3.13

▼ Link: Trigonometry in mathematics

■ ATL

■ **Transfer skills**: Make connections between subject groups and disciplines

You will have noticed that the resultant depends not only on the size but also on the relative directions of the forces – that is, on the angles they make.

We can calculate the size of the 'effect' a force has in a certain direction by using trigonometry. Think of a force making a triangle with the x- and y- axes as shown in Figure 3.14.

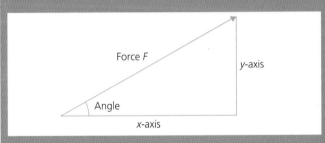

■ **Figure 3.14**

Then label the sides relative to the angle as 'opposite' and 'adjacent.' The longest side has a special name: it is called the hypotenuse.

The ratios of these sides are then defined as:

$$\text{sine } (angle) = \frac{\text{opposite side}}{\text{hypotenuse}}$$

$$\text{cosine } (angle) = \frac{\text{adjacent side}}{\text{hypotenuse}}$$

Rearranging,

opposite side = hypotenuse × sine (angle)

adjacent side = hypotenuse × cosine (angle)

This means that, if we know the length of the hypotenuse and the ratios (sine or cosine), we can work out the lengths of the sides of the triangle. See Figure 3.15. If our triangle is formed from a force, then

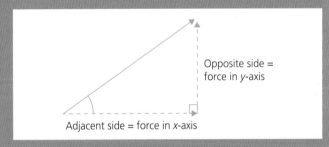

■ **Figure 3.15**

opposite side = how much of the force is pulling in the y-axis direction

adjacent side = how much of the force is pulling in the x-axis direction

For example, if the angle = 30°, then

sine 30 = 0.5

cosine 30 = 0.9

A force of 1 N at 30° therefore pulls with approximately 0.9 N in the x-axis direction, and approximately 0.5 N in the y-axis direction!

How do forces affect form?

STRETCHING THE POINT?

Have you ever done a bungee jump?

When we apply force to materials, they change shape by different amounts. We call the force that causes the change a **load**, and the change of shape is called **deformation**.

If the material returns to its original shape after the force is removed, it is said to be **elastic**. If – on the other hand – the material does not return to its original shape and dimensions even after the force is removed, it is said to be **plastic**.

ACTIVITY: Plastic or elastic?

> ■ ATL
>
> ■ **Critical-thinking skills**: Test generalizations and conclusions

Some students have made predictions about which of some materials are plastic, and which are elastic. They haven't tested their predictions, however.

Think about the behaviour you would expect if the materials were plastic, or if they were elastic.

Pair with a partner and **discuss**; are the students' predictions correct?

In your pairs, test the materials accordingly. Which variables will you **control**, and what will you observe?

Share your results with the class.

Plastic materials?	Elastic materials?
modelling clay	muffin
powerball	balloon
butter	football

ACTIVITY: Weird materials

> ■ ATL
>
> ■ **Information literacy skills**: Access information to be informed and inform others

■ **Figure 3.16** A bowl of corn starch on a loudspeaker. The sound waves cause the non-Newtonian fluid to go rigid

Some materials can't decide whether to be elastic or to be plastic.

Try making a solution of corn starch, and pouring it into a tray.

Slowly, push your finger into the corn starch. Now, 'jab' or 'prod' the fluid quickly. Notice anything?

Watch this video clip of some people running across a vat of corn starch (yes, it's possible!):
http://youtu.be/f2XQ97XHjVw

Corn starch is an example of a non-Newtonian fluid. The deformation of these fluids does not behave elastically, nor plastically.

Research to find out about other non-Newtonian substances!

ACTIVITY: Measuring deformation of a bungee elastic

Bungee rope must be **elastic**. What would happen if the bungee rope was **plastic**?

Watch a video here about testing bungee ropes to make sure they are safe: http://video.nationalgeographic.com/video/science/weird-science-sci/idkt-bungee-jump-testing/

Use the **experiment investigation cycle** in Figure 3.17 to design an experiment to find out how much elastic materials deform under load.

Safety: Make sure you check your design with your teacher for safety before starting.

◆ Assessment opportunities

This activity can be assessed using Criterion B: Inquiring and designing, and Criterion C: Processing and evaluating.

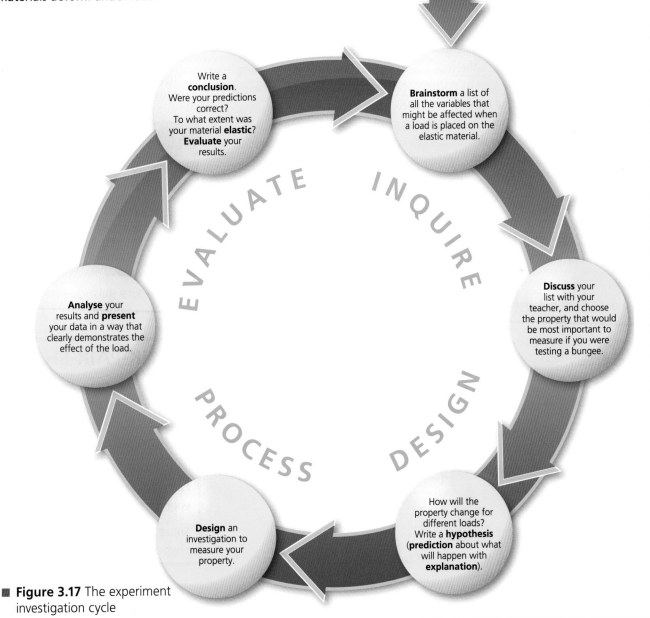

Figure 3.17 The experiment investigation cycle

Why do things balance?

BALANCING ACTS

Try these scientific magic tricks to impress your friends and family!

1 Take a large cardboard box. Find something quite heavy – maybe 1 or 2 kilogrammes in weight – and tape it securely into one corner of the box. For your magic trick walk into the room carrying your 'heavy box', grunting and groaning from the weight of it. Then, rest the corner where the masses are taped on the edge of a table (Figure 3.18).

Now remove your hands. … The box stays up! Magic!

2 Take a broom or something similar. Balance it on the index fingers of each of your hands at each end of the broom. Now slowly slide your fingers in towards the centre of the broom.

The broom will always balance and your fingers will always finish at the same place!

The balancing point of an object is called the centre of gravity (sometimes, centre of mass, which is the same place!)

When an object is supported at one point (a pivot or fulcrum), its weight is spread out or distributed around that point. As long as the mass of the object is distributed evenly – so it doesn't get fatter or thinner – we can imagine it behaving as though the weight w is concentrated at each end of the object (see Figure 3.20).

To work out whether an object will balance, we have to find the turning effect, or moment of the forces acting around the fulcrum. This is calculated by:

Moment (N m) = force (N) × distance from fulcrum (m)

■ **Figure 3.20** Turning forces on a balanced beam

■ **Figure 3.18** Balancing a box with a hidden weight

■ **Figure 3.19** Balancing a broom on the index fingers of each hand

THINK-PAIR-SHARE

On your own: Draw force diagrams for each of the two examples above, showing how the forces act on the box or the broom.

Pair: Compare your diagrams with your partner.

Share: Explain to each other the physics behind the 'tricks'.

ACTIVITY: Finding the balance point

In this activity you will find the centre of gravity for some two-dimensional objects.

You will need:
- **Stiff card**
- **Lab stand and clamp**
- **Drawing pin/thumb tack**
- **Thread**
- **A bung or cork**
- **Small mass, e.g. piece of modelling clay**
- **Felt-tip pen**

■ Figure 3.21

On the card draw a shape of any kind. Cut out your shape.

Now use the lab stand to clamp the bung/cork above your work surface.

Take the thumb tack and use it to pin your shape into the bung/cork. Make sure that the shape can turn easily under its own weight.

Tie the thread to the thumb tack and then attach the small mass to the other end of the thread. Let the thread fall vertically.

Use the felt-tip pen to draw along the line of the thread.

Now take out the thumb tack, and turn your shape around. Pin it again to the bung, in a different position.

Draw along the line the thread makes again. Repeat at least two more times.

You should find that the lines you draw all intersect at one place on the card. This is the **centre of gravity** for your shape.

To test, take the shape and see if you can balance it on the end of your finger at the centre of gravity!

Write a conclusion: How does this experiment work? Why do the objects balance where the lines cross?

◆ Assessment opportunities

This activity can be used to practise your ability to interpret observations and draw conclusions. This is important for Criterion C: Processing and evaluating.

Example: is the following balanced?

Figure 3.22 shows the following:

  moment on left side = 10N × 1.0m

  moment on right side = 20N × 0.5m

    moment (left) = 10Nm

  moment (right) = 10Nm

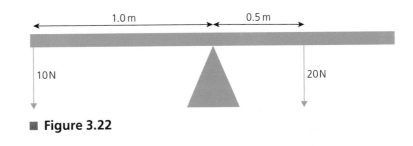

■ Figure 3.22

So, the beam will balance, as the turning effect on each side of the fulcrum is the same.

EXTENSION: THINK–PAIR–SHARE

■ ATL

- **Creative-thinking skills:** Apply existing knowledge to generate new ideas, products or processes

Discuss: Where is the centre of gravity of a doughnut? Is there more than one possibility? What other shapes might have similar properties?

■ Figure 3.23 A doughnut

■ Figure 3.24 Aircraft are designed so that their centre of gravity (C of G) lies under the wings. The whole aircraft then balances around this point, like a large flying 'lever'!

What is the relationship between structural form and strength?

STRUCTURE, FORCE AND FORM

The six pictures in Figure 3.25 show some human-made and some naturally occurring structures. Look carefully at the structures, and their **form**.

Can you pair the pictures according to the form, or shapes, they share?

■ **Figure 3.25**
1 Arches National Park, Utah, USA
2 Prairie grass seed viewed under scanning electron microscope
3 Burj Khalifa tower, Dubai, United Arab Emirates
4 Termite mound, Northern Territory, Australia
5 Sydney Harbour Bridge, Australia
6 Soccer ball

We have learned from nature that some structural shapes are stronger than others. For example, if we want to make a bridge across a gap, the simplest possibility is to lay a beam or bar across the gap (see Figure 3.26).

When load is applied to the centre of the beam, how will the beam deform?

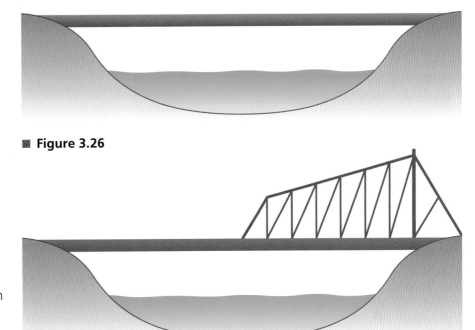

■ **Figure 3.26**

The problem of deformation in beam bridges could be solved by adding extra supports underneath the centre of the bridge. This might be difficult though, if we were trying to bridge a very deep chasm, or if we actually wanted things to walk under the beam – as in a doorway.

Another way to solve the problem of beam deformation is to add support above it.

This arrangement is called a **cantilever** (Figure 3.27).

■ **Figure 3.27**

And still another possibility is to bend the beam to provide support at either end.

This arrangement is called an **arch** (Figure 3.28).

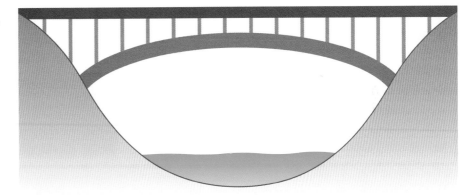

■ **Figure 3.28**

ACTIVITY: Predicting structural deformation

ATL

■ **Critical-thinking skills**: Use models and simulations to explore complex systems and issues.

Sketch copies of each of the bridge structures above, and add a load in the centre.

Then draw force arrows for:
● the load
● the reaction of the bridge to the load.

Now draw arrows to show how each of the bridges might transfer the force of the load into the sides of the valley.

Why do some buildings stay standing longer than others?

ACTIVITY: Measuring the strength of structural forms

■ ATL

- **Critical-thinking skills**: Use models and simulations to explore complex systems and issues

In this activity you will try out different structural forms to see which can carry the greatest load.

You will need:
- **Lots of strips of thin card, approximately 2 cm × 30 cm**
- **A paper glue-stick**
- **Small masses (10 g approximately)**
- **A set-square**

Method

Use the strips of card to construct structural shapes that will span a 20 cm 'gap', as follows:
- **a simple beam bridge**
- **a cantilevered bridge**
- **an arch.**

Hypothesis

Which of the structural shapes will be the strongest? How will your experiment show this? Explain your hypothesis with reference to the activity on page 51, and what we have learned about forces and structural forms.

Variables

To control: Use only single thicknesses of card for each part of your structure. (Why?)

Make the structures the same approximate dimensions. (Why?)

To change: Once you have built your structures, place 10 g masses at the centre of each span. Increase the mass each time.

To measure: Decide the best way to measure the deformation of the centre of the span for each structure.

Data analysis

Record your data in a suitable format, then present it so that the structural shapes can be compared.

Conclusion

Write a conclusion explaining how the structures supported the load in each case. Was your hypothesis correct?

Evaluation

Was your data sufficiently reliable to give you a fair comparison between the structures?

◆ Assessment opportunities

This activity can be assessed using Criterion C: Processing and evaluating.

ACTIVITY: Give me shelter

! Take action

- **! Find out about emergency aid.**

A number of international aid organizations such as Shelter Box and Oxfam provide easy-to-assemble, low-cost shelters for people who are displaced in disaster zones.

Use emergency shelters to find out what structures are used to meet these requirements, and how modern materials have helped them do so more effectively.

Write a design brief for a disaster aid organization, in which you **analyse** one such shelter design and **explain** how the physics of force and structure make it work.

Evaluate the costs, advantages and any disadvantages of the shelter, and **discuss** these in your report.

◆ Assessment opportunities

This activity can be assessed using Criterion D: Reflecting on the impacts of science.

SOME SUMMATIVE PROBLEMS TO TRY

Use these problems to apply and extend your learning in this chapter. The problems are designed so that you can evaluate your learning at different levels of achievement in Criterion A: Knowledge and understanding.

THIS PROBLEM CAN BE USED TO EVALUATE YOUR LEARNING IN CRITERION A TO LEVEL 3–4

1 A sailing boat is being blown by a wind, force *A*. At the same time, the captain is driving the boat with the engine, which produces the force *B*.

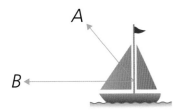

■ **Figure 3.29**

a **Outline** the effect of each of the forces on the motion of the boat.

b Using force arrows, show the direction of the resultant force on the sailing boat.

c **Calculate** the size of the resultant force on the sailing boat.
Scale: 1 cm = 500 N

THIS PROBLEM CAN BE USED TO EVALUATE YOUR LEARNING IN CRITERION A TO LEVEL 3–4

2 A skier is skiing downhill.

■ **Figure 3.30**

a Copy the forces in the diagram, then **label** the weight, *W*, of the skier and the reaction, *R*, of the slope.

b **Show** the direction of the resultant force on the skier.

c **Calculate** the size of the resultant force on the skier.

The skier accidentally skis onto a patch of grass and she experiences a force caused by friction with the grass. This force stops the skier from moving.

d On your diagram, **draw** an arrow showing the direction and estimated size of this force.
Scale: 1 cm = 400 N

THIS PROBLEM CAN BE USED TO EVALUATE YOUR LEARNING IN CRITERION A TO LEVEL 5–6

3 Pietro has secured a flagpole with two ropes, **A** and **B**. However, the arrangement makes the flagpole bend over in one direction. The diagram below shows the forces exerted on the flagpole by the ropes A and B.

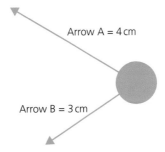

Arrow A = 4 cm

Arrow B = 3 cm

■ **Figure 3.31**

a On a diagram, **draw** a force arrow to show how Pietro should position a third rope, **C**, in order to make sure the flagpole is stable.

b Is there a better way to position all three ropes, **A**, **B**, **C**? **Explain** your answer.
Scale: 1 cm = 200 N

Figure 3.33 Millau viaduct, France

Figure 3.34 The Iron Bridge, Shropshire, UK

The **tensile strength** of a material is the maximum load that a material can support without fracture when being stretched, divided by the original cross-sectional area of the material.

The **compressive strength** is the capacity of a material or structure to withstand loads tending to reduce size, also expressed as the force per cross-sectional area.

a   What designs have been used for each of the bridges?

b   For each of the bridge shapes, **sketch** force arrows to show how a load placed at the centre of a span would be distributed or 'spread out' by the structure. **Show** on your diagrams where:
   ■ the structure is in **tension**, or being stretched
   ■ the structure is in **compression**, or being squashed.

c   One steel cable of the Millau viaduct is made from an average of seven strands of steel cable, and each strand is 1 cm in diameter. **Calculate**
   i   the load required to break one of the steel cables
   ii the compressive strength of one of the steel cables.

d   One strut (beam) of the Iron Bridge is around 0.25 m × 0.25 m. **Calculate**
   i   the load required to break one of the struts
   ii the compressive strength of one of the struts.

e   With reference to your calculations and to the force diagrams you drew in **b** above, **evaluate** how the form of the Iron Bridge and of the Millau viaduct is determined by the materials used in each.

THIS PROBLEM CAN BE USED TO EVALUATE YOUR LEARNING IN CRITERION A TO LEVEL 7–8

4   A shop sign is hung from a brick wall with a single post, **P**. The weight of the sign, *W*, can be considered to act down from the end of the post as shown in the diagram.
   The dotted lines **A**, **B** and **C** show three possible positions in which to put an additional chain to make the sign more secure.

   a   **Analyse** the diagram and **state** which is the best position for the chain, **A**, **B** or **C**?

   b   Why do you think this? **Explain** your answer.

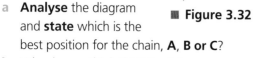

Figure 3.32

5   Figure 3.33 shows the Millau viaduct in south-western France. The viaduct was designed by the structural engineer Michel Virlogeux and Norman Foster Associates, in 2004.
   Figure 3.4 shows the Iron Bridge in Shropshire, England. The bridge was completed in 1779.
   The table below gives some specifications for the different bridges.

Bridge	Materials	Material tensile strength (N mm⁻²)	Material compressive strength (N mm⁻²)	Longest single span (m)	Highest clearance to ground level (m)
Millau viaduct	steel cables steel span	550	310	342	270
Iron Bridge	cast iron	345	690	30.6	16.75

Reflection

In this chapter we have analysed complex systems of forces by identifying and isolating individual forces, then calculating their combined effect. We have then been able to understand how forces balance in a state of **equilibrium**. Finally, we considered how structures observed in nature have helped us to build our own stable structures.

Use this table to reflect on your own learning in this chapter.		
Questions we asked	Answers we found	Any further questions now?
Factual: What are the effects of forces? What are the strongest shapes?		
Conceptual: How do forces work together? How do forces affect form? Why do things balance? What is the relationship between structural form and strength? Why do some buildings stay standing longer than others?		
Debatable: Are all the solutions to humanity's problems already to be found in nature?		

Approaches to learning you used in this chapter	Description – what new skills did you learn?	How well did you master the skills?			
		Novice	Learner	Practitioner	Expert
Creative-thinking skills					
Critical-thinking skills					
Transfer skills					
Information literacy skills					
Learner profile attribute(s)	Reflect on the importance of caring for our learning in this chapter.				
Caring					

4 How far, how fast, how much faster?

○ To know where we are and where we are **moving** to we need to describe the **relationship** between **space and time**.

■ **Figure 4.1** Olympic medallist athlete Mark Lewis-Francis preparing to race a BMW sports car

CONSIDER AND ANSWER THESE QUESTIONS:

Factual: How do we locate objects? How do we measure motion?

Conceptual: How can we present motion? What kinds of motion are there?

Debatable: To what extent is the world becoming a smaller place?

Now **share and compare** your thoughts and ideas with your partner, or with the whole class.

○ IN THIS CHAPTER, WE WILL ...

■ **Find out** how motion can be measured, and how it can be changed.
■ **Explore:**
  ■ how humans first circumnavigated the world, and how much more quickly we could do it now
  ■ whether athletes can outrun sports cars and how it is possible to accelerate even when speed does not change
  ■ how ideas about motion have changed, and what we can learn from the way they have changed.
■ **Take action** to raise awareness about the dangers of speeding traffic and evaluate the adequacy of speed controls where you live.

MAN VS MACHINE

In a publicity stunt that took place in 2012, the German car manufacturer BMW staged a race between the British Olympian Mark Lewis-Francis and one of its fastest production sports vehicles.

Who would you predict to win this race?

Watch the video of the test by doing an internet search for Lewis-Francis vs BMW video. While watching, listen carefully to the commentary by the sports scientist, and make note of the factors affecting the performance of the car and the athlete in the race.

THINK-PAIR-SHARE

Were you right? Or were you partially right? Under what **conditions** was your prediction correct?

KEY WORDS

distance	speed
displacement	velocity

■ These Approaches to Learning (ATL) skills will be useful …

- Collaboration skills
- Information literacy skills
- Critical-thinking skills
- Transfer skills

◆ Assessment opportunities in this chapter

- ◆ **Criterion A**: Knowing and understanding
- ◆ **Criterion B**: Inquiring and designing
- ◆ **Criterion C**: Processing and evaluating
- ◆ **Criterion D**: Reflecting on the impacts of science

● We will reflect on this learner profile attribute …

- ● Knowledgeable – using concepts; exploring the way scientific knowledge can help understanding in other subjects; engaging with issues that have local and global significance.

How do we locate objects? How do we measure motion?

ACTIVITY:
Circumnavigating the world

■ **ATL**

- **Information literacy:** Access information to be informed and inform others
- **Critical-thinking skills:** Interpret data
- **Transfer skills:** Make connections between subjects and disciplines

Inquiry: How far did Magellan's fleet travel? How fast did they go?

The table shows the main stopping-off points on Magellan's voyage. Note that not all dates are known for all places because Magellan's original log books were lost.

HOW FAR?

The first successful circumnavigation of the world is usually attributed to the Portuguese explorer Ferdinand Magellan, who was commissioned by King Carlos V of Spain to find a route to the Spice Islands (Las Moluccas, now Maluku Islands) of the south-western Pacific. This widely held historical 'fact' is actually inaccurate: Magellan certainly had the idea for the trip and put the whole adventure together, but was killed towards the end of the journey in East Timor, when he tried to overwhelm an indigenous people. His Spanish Basque navigator Juan Sebastián de Elcano commanded the ships' return, with only 18 of the original 270 crew surviving.

The map (Figure 4.2) shows the route Magellan's fleet took, starting and ending in the port of Seville in Spain.

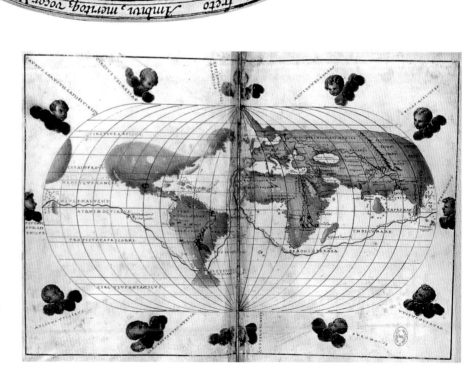

■ **Figure 4.2** 1544 map showing the route taken around the world by the Portuguese explorer Ferdinand Magellan, between 1519–1522

Place	Latitude, Longitude	Date of departure	Distance (km)
Seville, Spain	37°N 6°W	August 10, 1519	0
Canary Islands	28°N 16°W	September 1, 1519	
Brazil, Tropic of Capricorn	45°W	December 13, 1519	
Strait of Magellan	52.5°S 69°W	October 21, 1520	
Entered Pacific Ocean		November 28, 1520	
Mariana Islands	18°N 145°E	March 6, 1521	
Philippines	10°N 123°E	March 16, 1521	
East Timor	9°S 126°E	December 21, 1521	
Cape of Good Hope	35°S 20°E	March 20, 1522	
Cape Verde Islands	16°N 24°W		
Seville, Spain	37°N 6°W	September 10, 1522	

1 Use an online distance calculator to work out the distances between points, and so the total distance travelled in the voyage.
Search online using a global distance calculator. Take care to find a calculator that works in **metres** or **kilometres**.

THINK-PAIR-SHARE

If you were flying rather than sailing, what is the distance around the Earth's equator?

2 **Analyse** the data for the Magellan fleet's voyage.
   - At the end of the voyage, how far had Magellan travelled?
   - How many times did Magellan travel the Earth's equatorial circumference?
3 **Evaluate** the validity and the reliability of the historical data. To what extent does it enable us to answer the inquiry question? What are the limitations of the data? What other information would we need?

▼ Links: Individuals and societies, history, economics

◼ ATL

■ **Transfer skills:** Inquire in different contexts to gain a different perspective

While Magellan and other great explorers are credited with real scientific achievements such as circumnavigating the globe, their motives were rarely scientific.

What motivated Magellan to leave Seville in the first place? How often is scientific or technological discovery achieved 'just to find out,' and how often are the motives economic?

Of course, Magellan didn't have to go to such lengths purely to circumnavigate the globe. Theoretically, he could have travelled all around the world in just a few steps (Figure 4.3).

Any journey which starts and ends at the same point demonstrates to us the need to distinguish between how far we have actually **moved**, and how far we have **travelled**. In physics, these measurements are called **displacement** and **distance** respectively.

◼ Table 4.1

Quantity	SI unit	Description	Variable letter(s)
distance	metres	how far we have travelled (scalar)	s
displacement	metres	how far we have moved overall; distance between starting point and ending point; vector	s, $\vec{s}$

Notice, then, that distance is just a number or scalar value, while displacement also includes the **direction** in which we moved – so it is a **vector** (see Chapter 3 for discussion of vectors and forces).

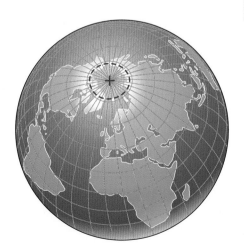

■ **Figure 4.3** Circumnavigating the world in just a few steps!

How do we measure motion? How can we present motion?

There have been many other attempts to circumnavigate the world since Magellan's. For example, in 2006, Tim Harvey undertook a trip from and to Vancouver using only means of transport that did not produce carbon emissions: **www.vancouvertovancouver.com/**

HOW FAST?

The quantity of speed relates the concept of movement to time: how far has an object's position changed in a certain amount of time? But it also sometimes matters **which way we are moving** – so we need a vector equivalent of speed (see Table 4.2).

Again, notice that **velocity** includes both the speed and the direction in which the object is travelling – so two bits of information about the motion. Notice too that we have defined speed and velocity as **average** values – we will come to this shortly, but perhaps you can already speculate why this might be?

■ Table 4.2

Quantity	SI unit	Description	Equation
average **speed**	metres per second = $m\,s^{-1}$	how far we have travelled in a certain time (scalar)	Average speed $v\ (m\,s^{-1}) = \dfrac{s\ (m)}{t\ (s)}$
average **velocity**	metres per second = $m\,s^{-1}$	how far we have moved between two points in a certain time and in a certain direction (vector)	$\boldsymbol{v}\ (m\,s^{-1}) = \dfrac{\boldsymbol{s}\ (m)}{t\ (s)}$

ACTIVITY: Magellan's speeds and velocities

Inquiry: How far did Magellan's fleet travel? How fast did they go?

■ ATL

■ **Critical-thinking skills:** Interpret data; Recognize unstated assumptions and bias

Use the equations and the data you gathered earlier for Magellan's voyage to **estimate**:

● the average speed for the entire voyage

> Think carefully about the units of your data and, so, of your answer!

● the average speed over the part of the voyage that took the longest
● the velocity for the entire voyage.

In 1967, Sir Francis Chichester circumnavigated the globe alone in his yacht, *Gypsy Moth IV.* He started and ended in Plymouth, England, and the voyage took 226 days. Because he passed through two 'antipodean points' on the route, we know for sure that the distance he travelled was equivalent to a circumnavigation of the equator.

Compare the information we have for Sir Francis Chichester's route and for the Magellan voyage. **Evaluate** the reliability and the validity of the data for each. Why must we refer to average speeds and velocities in the case of either voyage?

◆ Assessment opportunities

In this activity you have practised skills that are assessed using Criterion A: Knowing and understanding.

ACTIVITY: Measuring traffic speed

Inquiry: how well do speed restrictions work?

Is there a traffic camera near to your school? Or a speed restriction sign? If so, you can try this experiment. If you don't have a traffic camera or other speed restriction near to your school, you can still use this method to measure traffic speeds along a busy stretch of road. Are there signposts telling drivers where your school is located? Is there any other hazard warning? If so, use the method to see what effect the warnings have on driver behaviour.

You will need to work in a large group of 8–10 people. You should read the following instructions and discuss the points at the end of the procedures before you start.

Equipment
- **Stopwatch or other accurate timer**
- **Long tape measure (10–20 m minimum)**

Procedure 1
One member of your group will be the **timer**. They will also need two assistants: an **observer** and a **recorder** to write down times.

Position the other members of your group along the pavement (sidewalk) before and after the traffic camera, at regular intervals, for example every 10 m or so. Measure the intervals with the tape measure.

Marker 1 is the person that the traffic passes **first**. Marker 1 identifies a particular vehicle and, when it passes them, raises their hand.

When Marker 1 raises their hand, the **observer** says 'start!' and the **timer** starts the stopwatch.

As the vehicle passes each of the other markers in the group, they raise their hands. As they do so, the observer counts them off, saying '1... 2... 3...' as the vehicle passes each of the markers.

The **timer** reads out the time for each marker; the **recorder** writes the time in a data table.

Repeat for a suitably large number of vehicles.

Procedure 2: The rainy day option
You can observe traffic through traffic cameras all over the United States here: http://trafficland.com/

You will need to estimate distances in the images however, perhaps using easily recognizable objects such as lamp-posts, etc.

Before you begin, discuss these questions:
- **How many vehicles will be a suitable number?**
- **How will you organize the table for the recording of the raw data?**

Hypothesis
Now, individually, make a hypothesis about the way in which the traffic will behave in the area you are monitoring for the experiment. Be sure to refer to your experiment variables, for example: 'I think that the times measured for the traffic will .... because the traffic will ...'

PRESENTING JOURNEYS

Traffic police officers make sure that we stick to the rules when it comes to motion. Modern cars can easily exceed even the maximum speed limits for many countries. But why make them so fast? How safe is it to drive at high speed?

Figure 4.4 Traffic police officers use radar guns to measure the speed of traffic

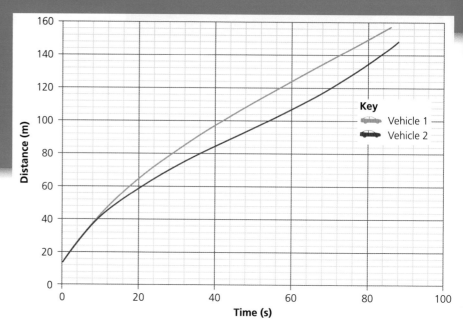

■ **Figure 4.5** Distance travelled by two vehicles with time

One way to present a journey is to graph distance against time, as shown in Figure 4.5, for example.

Which vehicle was travelling most quickly? To be sure about the answer to this question, we could read off the distance travelled by each vehicle between two given times. Then,

$$\text{average speed} = \frac{\text{distance travelled between two times}}{\text{time 2} - \text{time 1}}$$

For example, from the graph, we can read off that at 40 s Vehicle 1 has travelled 96 m. At 60 s Vehicle 1 has travelled 124 m. So

$$\text{average speed} = \frac{124 - 96}{60 - 40}$$

$$\text{average speed} = \frac{28}{20} = 1.4\,\text{m s}^{-1}$$

If we then calculate the speed of the other vehicles at the same time, we can show the journey in terms of the speeds of the vehicles, rather than in terms of the distances they have travelled (see Figure 4.6).

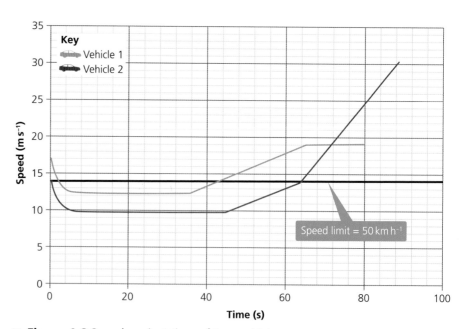

■ **Figure 4.6** Speed against time of two vehicles

1 Use your raw data to **calculate** the speeds of the vehicles at each of the measurement points.
2 **Present** this information in your own graph of speed against time for all vehicles.
3 **Discuss** and then **outline** the information in your own graph of speed against time for all vehicles:
- Which vehicle changed speed most quickly? **Explain** how you decided.
- **Interpret** your graph. **Identify** any times where the vehicles travel at constant speed.
- **Calculate** the speed limit for the road in m s⁻¹, and mark this on the graph with a line – as shown in Figure 4.6 for 50 km h⁻¹. Remember, you will need to convert the speed limit from the units used where you live to the standard scientific unit of m s⁻¹.

> **Hint**
>
> $1\,\text{m s}^{-1} = 0.001 \times 60 \times 60 = 3.6\,\text{km h}^{-1}$
>
> $1\,\text{mile per hour (mph)} = 1.61\,\text{km h}^{-1}$

- Use a vertical line to show on your graph the position of the traffic camera, speed restriction sign or similar.

Conclusion

Summarize your findings, referring closely to your data. Explain how you have interpreted the data to give information about the motion of the vehicles. Describe the effect of the speed camera or other restriction on the behaviour of the drivers.

Evaluation

Outline any problems (sources of error) in your experiment.

Evaluate the importance (significance) of each of these problems.

Suggest how you could modify the design of your experiment to remove or lessen these problems.

◆ Assessment opportunities

This activity can be assessed using Criterion C: Processing and evaluating.

What happens to the **slope** of the distance–time graph as the vehicles speed up or slow down?

If we think back to our best-fit line technique in Chapter 2,

$$\text{gradient of the graph } \frac{\Delta y}{\Delta x} = \frac{\text{change in } y}{\text{change in } x} = \frac{\text{change in distance}}{\text{time}}$$

This means the gradient (slope) on a distance–time graph is the **average speed** for that part of the journey.

There is an important nuance here. Notice that we have to calculate the gradient between two points on the graph; this means that we are always assuming the speed **stays the same** between those two points, so it is an **average speed**.

▼ Link: Mathematics

 ATL

- **Transfer skills**: Make connections between subject groups and disciplines

To know the speed at one particular point, we would have to calculate the change in distance over a very tiny change in time, in fact over a change in time that was vanishingly small. This speed is called the **instantaneous velocity** and to work it out we would need to use a mathematical process called **calculus**, which was invented by Sir Isaac Newton ... more of him later.

What about the speed–time graph? In this case,

$$\text{gradient of the graph } \frac{\Delta y}{\Delta x} = \frac{\text{change in } y}{\text{change in } x} = \frac{\text{change in speed}}{\text{time}}$$

The quantity

$$\frac{\text{change in speed}}{\text{time}}$$

is called the **acceleration**.

It has units of

$$\frac{\text{m s}^{-1}}{\text{s}}$$

or 'metres per second per second' – which we write more simply as $\mathbf{m\,s^{-2}}$.

Acceleration can be positive or negative in value. If the **final velocity** – usually written as *v* – of the object is lower than the initial velocity – usually written as *u* – then acceleration is a **negative** number.

A negative acceleration for an object slowing down is sometimes called a **deceleration** or more correctly a **retardation** in velocity. Because acceleration must always have a sign, it always has a **direction** and so is always a **vector**.

Table 4.3

Quantity	S.I unit	Description	Equation
acceleration	metres per second per second = m s^{-2}	how much velocity has changed in a second (vector)	acceleration $a\ (\text{m s}^{-2}) = \frac{(v - u)\ (\text{m s}^{-1})}{t\ (\text{s})}$

The gradient (slope) on a speed–time graph is the **acceleration** for that part of the journey.

ACTIVITY: Analysing accelerations

ATL

- **Critical-thinking skills:** Recognize and evaluate propositions

Using the graphs from your 'traffic speeds' activity, **think-pair-share** the following points. **Outline** your answers in words and diagrams, as appropriate.

1 What is the acceleration when the velocity of the vehicle is constant? **Describe** what this looks like on the speed–time graph.

2 What is the effect on the curve on a speed–time graph when the vehicle speeds up but then hits the brakes? **Describe** or sketch what this would look like on the speed–time graph.

3 Since acceleration is a vector, direction matters. **Explain** under what circumstances it is possible for an object to accelerate without changing speed.

Assessment opportunities

In this activity you have practised skills that are assessed using Criterion A: Knowing and understanding.

ACTIVITY: Staying alive on the roads

Take action

! Think about road safety.

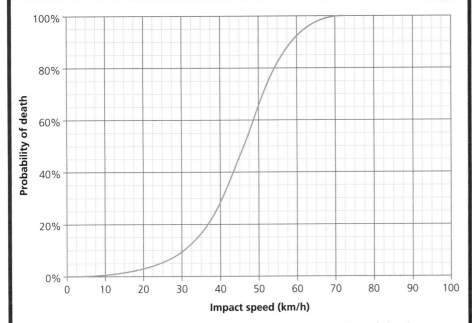

Figure 4.7 Graph showing relationship between probability of death and impact speed – source: Improving Pedestrian Safety (Curtin-Monash Accident Research centre)

Look at Figure 4.7.

Discuss: What is the relationship between speeding and traffic mortalities? Could more be done in your locality to restrict or control speeding? Write a report to the local authority, in which you:

- choose one form of speed-limiting device. **Explain** how the device works.
- **evaluate** the effectiveness of the speed-restriction methods that are in place.
- **outline** how your chosen speed-limiting device might affect speeding in your area.
- make sure that you support your arguments with supporting data from your own enquiries and reference any research you have done.

Assessment opportunities

This activity can be assessed using Criterion D: Reflecting on the impacts of science.

e **Interpret** your graph to show the regions of:
- minimum speed
- maximum speed
- maximum velocity
- minimum velocity.

f Using the data or your graph, **calculate** Mateo's maximum and minimum velocity.

SOME SUMMATIVE PROBLEMS TO TRY

Use these problems to apply and extend your learning in this chapter. The problems are designed so that you can evaluate your learning at different levels of achievement in Criterion A: Knowledge and understanding.

THIS PROBLEM CAN BE USED TO EVALUATE YOUR LEARNING IN CRITERION A TO LEVEL 3–4

1 The data in the table shows position measurements for Mateo's journey to school in the morning.

Mateo's action	Time (minutes)	Distance travelled (m)
home	0	0
walk to metro station	5	250
forgot physics book; walk home again to get it	5	250
walk to metro station	4	250
on metro train	2	1000
station	2	0
on metro train	2	1500
walk from metro station to school	4	200

a **Design** a new table to present the following calculated data:
- i the total time travelled
- ii the total distance travelled

b Taking home as zero, and assuming Mateo travels in a straight line to school, **calculate** the total displacement from home.

c **Outline** how the distance travelled by Mateo compares to the distance from home to school. **Explain** the difference in these values.

d **Plot** a graph to show distance and displacement for Mateo's journey on the same axes.

THIS PROBLEM CAN BE USED TO EVALUATE YOUR LEARNING IN CRITERION A TO LEVEL 5–6

2 The graphs below show the motion of a sports car and an athlete for the first 10 seconds of a race.

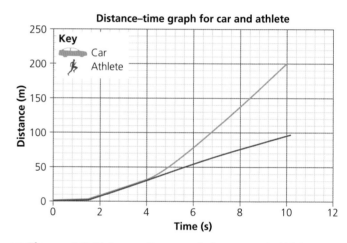

■ **Figure 4.8** Distance–time graph for car and athlete

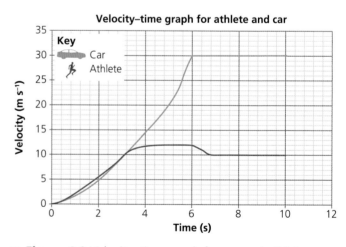

■ **Figure 4.9** Velocity–time graph for car and athlete

a **Interpret** the information in the graphs to find:
- i the maximum velocities achieved by both car and athlete

ii the distance over which the athlete is winning the race

iii the time at which the car is first travelling faster than the athlete.

b **Analyse** the information in the graphs to **calculate**:

i the average velocity of both car and athlete in the first 2, 4 and 6 seconds

ii the greatest acceleration achieved during the race. **State** at what time this was achieved, and by which racer (car or athlete).

c **Analyse** the information to **compare** how each of the racers performed in the first 5 and in the last 5 seconds of the race. **Explain** how the athlete can initially out-run the car, even though the car has a much greater maximum velocity.

THIS PROBLEM CAN BE USED TO EVALUATE YOUR LEARNING IN CRITERION A TO LEVEL 7–8

3 Flight MY100 is about to leave the terminal at your local airport and take off. The co-pilot is in charge of making sure that the aircraft's journey goes smoothly. Before take-off the aircraft 'taxis' away from the terminal, out across the airport to the runway. Then it takes off.

The co-pilot has three charts for this part of the journey. The first chart (Figure 4.10) is for the time taken for each distance the aircraft must travel from the terminal.

a **Outline** what is happening to the aircraft at the following times. **Justify** your answers with reasons.
- between 0 s and 70 s
- between 80 s and 160 s
- between 160 s and 200 s

Oh dear – in a rather embarrassing situation, the co-pilot has managed to mix up the remaining two charts. He knows that one of them shows the velocity the aircraft must have during each part of the taxi, and the other shows the acceleration the aircraft must have. These are the two graphs (Figures 4.11 and 4.12).

b **State** which graph is which. **Justify** your answer with reference to the motion of the aircraft.

c What is happening to the aircraft between 60 s and 80 s? **Outline** your answer using all of the words in the box:

> acceleration   time   velocity

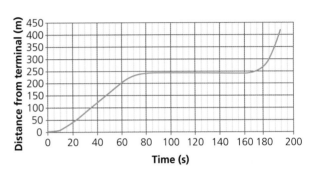

Figure 4.10 Flight MY100 distance travelled before take-off

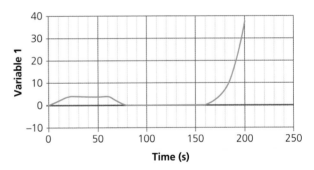

Figure 4.11 Flight MY100 taxi profile 1

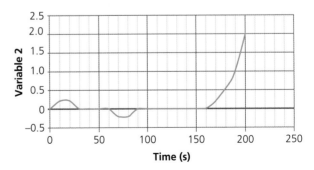

Figure 4.12 Flight MY100 taxi profile 2

Figure 4.13 shows the first part of the journey after MY100 has taken off.

Inside the cabin of the aircraft a small screen gives information about the flight for the passengers to read.

d Using information from Figure 4.13 **calculate** the missing information from the screen (Figure 4.14).

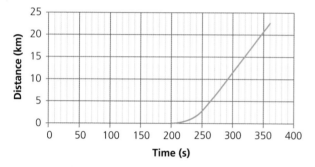

■ **Figure 4.13** Distance travelled after take-off for first part of flight MY100

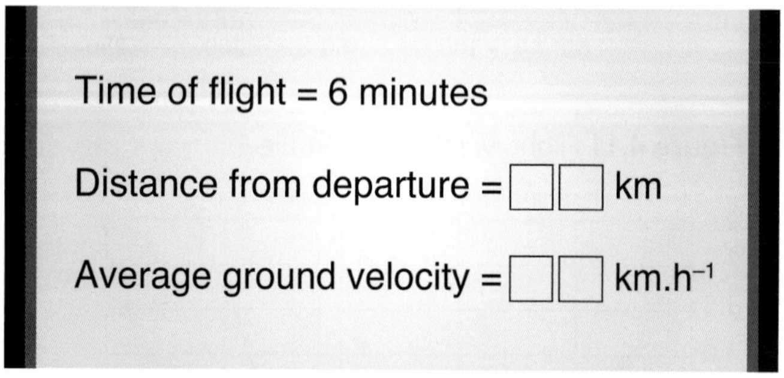

Time of flight = 6 minutes

Distance from departure = ☐☐ km

Average ground velocity = ☐☐ km.h⁻¹

■ **Figure 4.14** In-flight cabin display on a passenger aircraft

The co-pilot thinks that the instantaneous velocity at time = 6 mins is likely to be about the same as the average velocity for the whole flight. The pilot thinks that the instantaneous velocity at 6 mins is probably, in fact, more than the average velocity for the whole flight.

e **State** who you think is correct. **Explain** your answer with reference to your calculated value for average velocity and any other information from the graph.

f **State** over what range of times the co-pilot could be right that the instantaneous velocity is about the same as the average velocity.

g **Explain** your answer to **f** with reference to your calculated value for average velocity and any other information from the graph.

The velocity of the aircraft is measured using a device called a pitot, which is a small tube that samples the speed of the air flowing past the aircraft's body. This speed is called the air speed. The velocity of the aircraft relative to the ground is then called the ground speed. The ground speed is calculated using radar waves which reflect from the ground.

h If the aircraft is flying into a wind with speed 5 m s⁻¹, **state** and **explain** the effect this would this have on the velocity measured by the pitot.

The aircraft's ground speed radar has stopped working, and now the pilot has to rely on the measured air speed only.

i If the total journey for the aircraft was 1800 km long and the speed is as you calculated in **h**, **state** whether the flight would arrive late or early and **calculate** the difference in arrival time.

j Luckily the co-pilot receives a weather report from air-traffic control which includes the wind speed. **State** what the pilot should do to maintain the proper ground speed. **Summarize** the likely effect of the pilot's action on the fuel consumption of the aircraft, and on the environmental impact of the flight.

k It is often the case that flying west across the Atlantic Ocean (say, from Paris to New York) takes a longer time and more fuel than flying East. **Suggest** why this may be.

Reflection

In this chapter we have explored how place and time can be measured and represented, and how they are related by the quantities of velocity, speed and acceleration.

Use this table to reflect on your own learning in this chapter.		
Questions we asked	Answers we found	Any further questions now?
Factual: How do we locate objects? How do we measure motion?		
Conceptual: How can we present motion? What kinds of motion are there?		
Debatable: To what extent is the world becoming a smaller place?		

Approaches to learning you used in this chapter	Description – what new skills did you learn?	How well did you master the skills?			
		Novice	Learner	Practitioner	Expert
Collaboration skills					
Information literacy skills					
Critical-thinking skills					
Creative-thinking skills					
Transfer skills					
Learner profile attribute(s)	Reflect on the importance of being knowledgeable for our learning in this chapter.				
Knowledgeable					

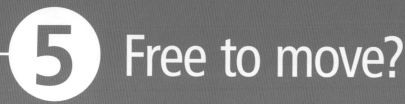

5 Free to move?

○ *Movement* is **change**, and our world has been changed by *freedom of movement*.

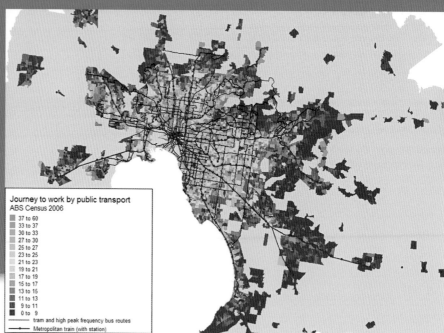

Journey to work by public transport
ABS Census 2006

37 to 60
33 to 37
30 to 33
27 to 30
25 to 27
23 to 25
21 to 23
19 to 21
17 to 19
15 to 17
13 to 15
11 to 13
9 to 11
0 to 9
—— tram and high peak frequency bus routes
—— Metropolitan train (with station)

■ **Figure 5.1** Daily travel by public transport in Melbourne, Australia

CONSIDER AND ANSWER THESE QUESTIONS:

Factual: What causes motion? What changes motion? What affects the motion of falling objects?

Conceptual: What relationship is there between force and motion?

Debatable: Does everybody have the right to travel the way they wish? Should everybody have the right to travel as they wish?

Now **share and compare** your thoughts and ideas with your partner, or with the whole class.

○ IN THIS CHAPTER WE WILL …

■ **Find out** the ways in which force causes change in motion, while matter prefers to carry on doing what it was doing before – **inertia**.

■ **Explore**:
  ■ how momentum can be used to quantify the motion of an object
  ■ how Newton's laws of motion help us understand and develop new forms of propulsion and transportation systems.

■ **Take action** to reflect on how access to transportation affects people's lives globally, and we will research and report on the debates around high-speed rail systems in particular.

■ These Approaches to Learning (ATL) skills will be useful …

■ **Communication skills**

■ **Critical-thinking skills**

■ **Collaboration skills**

■ **Creative-thinking skills**

■ **Information literacy skills**

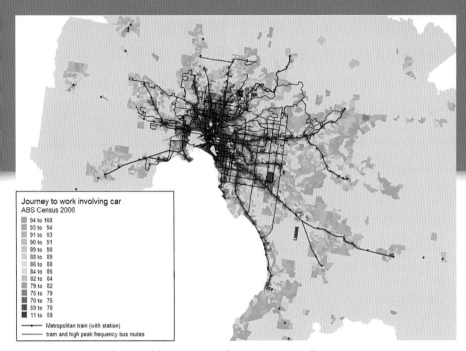

Journey to work involving car
ABS Census 2006

- 94 to 100
- 93 to 94
- 91 to 93
- 90 to 91
- 89 to 90
- 88 to 89
- 86 to 88
- 84 to 86
- 82 to 84
- 79 to 82
- 75 to 79
- 70 to 75
- 59 to 70
- 11 to 59

—•— Metropolitan train (with station)
——— tram and high peak frequency bus routes

■ **Figure 5.2** Daily travel by car in Melbourne, Australia

Look at Figures 5.1 and 5.2.

Figure 5.1 shows the number of journeys to work made by walking in the city of Melbourne, Australia. Figure 5.2 shows the number of journeys to work made by car for the same city. The blue lines show public transportation routes, such as tramways or railways.

Now consider this chart, which shows the mode of transport for people in Uganda during 2010 (Figure 5.3).

How are choices about transportation and movement likely to be different for people who live in urban Australia, and people in Uganda? Transportation and movement are key factors in promoting socioeconomic change. In this topic we will explore the physics of changes in movement caused by force. We will reflect on the different ideas that people have had over the millennia concerning the causes of motion, and we will look at the physics of different transportation systems. Finally, we will reflect on the impact of transportation modes on societies globally.

KEY WORDS

acceleration	mass
force	motion
inertia	momentum

● We will reflect on this learner profile attribute …

● Being open-minded – working with others means keeping an open mind about different points of view. Equally, to understand controversial issues, we need to consider both sides of an argument.

◆ Assessment opportunities in this chapter

◆ **Criterion A**: Knowing and understanding

◆ **Criterion B**: Inquiring and designing

◆ **Criterion C**: Processing and evaluating

◆ **Criterion D**: Reflecting on the impacts of science

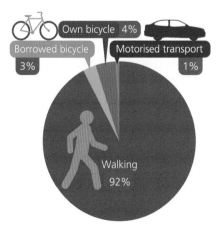

Own bicycle 4%
Borrowed bicycle 3%
Motorised transport 1%
Walking 92%

■ **Figure 5.3** Mode of transport Uganda 2010 – Data source: Institute for Transportation and Development Policy – Europe

What causes motion? What changes motion?

FORCE AND MOTION

Why do things move at all? You probably know the answer to this, but it has not always been obvious. For a very long time, the theory of motion of Aristotle was held to be the correct explanation for the mystery of movement. Look at Figure 5.4.

All that exists is made from four fundamental elements, mixed in various amounts: Earth, Water, Air and Fire. Each element has its natural place in the cosmos – Earth is lowest, for it invariably sinks, then Water above it, then the Air, and above all Fire, which rises. Objects move because of the elements they contain – they tend to their natural place in the cosmos. Thus, a stone held high in the Air will sink to the Earth, for a stone contains mostly Earth.

All objects, once in their natural place, will stop moving. When there is no reason to move, things do not move. All motion is caused therefore – when we cease pushing an object, it will cease to move.

■ **Figure 5.4** Aristotle (384–322 BCE)

We have already encountered Galileo Galilei, who gave us the idea of variables and the first experimental method. But he was also instrumental in changing our understanding of motion. Galileo criticized Aristotle's explanation of motion by using a simple thought experiment. He asked himself, what would happen if a stone were tied to a bladder (balloon) filled with air? What is the natural place for **this** object? Galileo proposed an alternative explanation for motion (Figure 5.5).

Matter has a tendency to inertia. It will continue in its current state, unless a force acts to change that state. Thus, an object that is stationary, in the absence of any force, will remain stationary. An object that is moving, in the absence of any force, will continue to move.

■ **Figure 5.5** Galileo Galilei (1564–1642)

ACTIVITY: To move or not to move?

In pairs, decide who is going to be an 'Aristotlean' and who a 'Galilean.'

Look at the following examples of actual motion. Individually, write:

● a **description** of what happens to the moving objects
● an **explanation** for the motion using the theory of motion you have chosen

1 A cannon ball after it leaves the cannon
2 A hollow wooden ball and an equally sized solid stone ball falling from a cliff top
3 The space probe *Voyager 1*, after its mission has finished in space with no rocket fuel remaining.

Now debate your explanations in class. Which theory works best with what we know about the motion of the objects?

Was Aristotle just crazy to believe that all objects would stop moving once force was no longer applied? Consider: What observations did he have available to him in making this theory? What do we now know causes objects on Earth to slow down and stop, that Aristotle did not know about? **Summarize** your ideas.

■ **Figure 5.6** A cannonball follows a curved (parabolic) trajectory. Inertia means it continues to move forward as the force of weight pulls it towards the Earth. What path for the cannonball would Aristotle's theory predict?

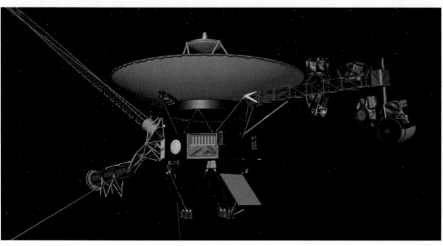

■ **Figure 5.7** The two *Voyager* space probes were launched in the mid-1970s. They are still travelling through deep space, although they ran out of propellant for their rocket engines many years ago. *Voyager 1* is the first human-made object to leave the Solar System.

> **First law of motion**
> An object at rest will remain at rest unless acted on by an unbalanced force. An object in motion continues in motion with the same speed and in the same direction unless acted upon by an unbalanced force.

■ **Figure 5.9**
Issaac Newton's three laws of motion

WHAT ARE THE LAWS OF MOTION?

Isaac Newton was – by coincidence – born during the same year that Galileo died, in 1642. Later in life, Newton had to flee Cambridge University following an outbreak of the plague. While taking refuge in his family manor at Woolsthorpe in Lincolnshire, England, he spent some time thinking about Galileo's ideas on motion. You may have heard the famous story about Newton snoozing under

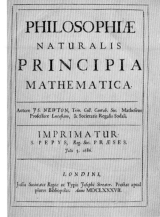

■ **Figure 5.8** Title page of Newton's first publication of *Principia Mathematica* (1686)

the apple tree, and observing an apple fall to Earth … although this is probably apocryphal (made up afterwards). Whatever inspired him, Newton had a 'miracle year' of ideas. Not only did he elaborate his own mathematical system for describing change (he called it the **calculus**), but he also reduced the mystery of motion to three laws which he later wrote in his *Principia Mathematica* (1686).

Newton's three laws of motion remained the foundation for most of physics for nearly 300 years, and are still a powerful tool for understanding most of the motion we observe in everyday life.

THE FIRST LAW OF MOTION

We can recognize in the first law Galileo's concept of **inertia**.

ACTIVITY: Inertia tricks

Here are some more physics magic tricks to demonstrate and amaze your friends!

1 **Inertia karate**
   Karate masters can smash very dense objects, such as wooden or concrete blocks, with their hands, provided the objects are suspended over a void. To get an idea how this is possible, try this:
   - Take a long stick of wood, no more than 1 cm diameter or 1 cm square, and around 1 m long.
   - Place the wood so that it balances on the rims of two plastic beakers.
   - Put on safety glasses! Make sure your audience is at least 2 m away.

- Using another stick – perhaps a broom handle or similar – whack the stick in the middle.
- Hey presto! If inertia has done its trick, you now have a broken stick and two intact beakers.

2 Place a plastic beaker full of water on the edge of a table. Place a sheet of paper beneath the beaker. Pull the sheet of paper **quickly** out from under the beaker! (A laboratory variation on the old 'table-cloth pulling' trick!)

3 Take a boiled egg and a raw egg. Spin both at the same time and the same speed. Watch what happens …

For each of the magic tricks, **outline** how inertia makes the unexpected happen.

> **Second law of motion**
> Acceleration is produced when a force acts on a mass. The greater the mass (of the object being accelerated) the greater the amount of force needed (to accelerate the object).

> **Third law of motion**
> For every action there is an equal and opposite reaction.

THE SECOND LAW OF MOTION

We have already seen in Chapter 2 how force can cause materials to deform (change shape). If the material is also mobile, then force will cause a change in its state of motion.

When a force is applied to an object for a certain amount of time, it changes the motion of the object. Newton called the 'push' that this force gave the impulse and deduced that it would be simply given as

$$\text{force (N)} \times \text{time (s)} = \text{change in momentum}$$
$$\Delta(mv)\,(\text{kg m s}^{-1})$$

Newton used the concept of **momentum** to describe the state of motion of an object, where momentum is given as

$$\text{momentum } (p) = \text{mass } (m) \times \text{velocity } (v)$$

When the momentum of an object changes, there are two quantities that can change: the force can change the **mass** of the object, and it can change its **velocity**. Since mass does not usually change spontaneously just because a force is applied (unless we break the object into bits or add something to it), then it must be the **velocity** which changes. Writing in our letter variables,

$$Ft = mv - mu = m(v - u)$$

Rearranging gives $F = m \dfrac{(v - u)}{t}$

The quantity $\dfrac{(v - u)}{t}$ is, of course, the **acceleration**,

so we can now write:

$$F = m a$$

So, a force acting on a mass produces an acceleration that depends on the size of the mass.

ACTIVITY: Measuring acceleration

> ### ■ ATL
>
> ■ **Critical-thinking skills:** Gather and organize relevant information to formulate an argument; Evaluate evidence and arguments

Inquiry: How is the acceleration of an object affected?

Aim: To measure the mass of a trolley by measuring its acceleration.

In this experiment you will use Newton's second law to measure the acceleration of a force on a rolling trolley, and so find the mass of the trolley (without weighing it).

Equipment
- A dynamics trolley or other rolling object
- A long ramp, length 1–2 m
- A pulley
- Some string
- Ten masses, 100 g or so each
- Ticker-tape timer, photo-optical timing gates, or ultrasonic distance-measuring device

Variables
- **Controlled:** mass of the whole system, M
- **Independent** (to change): accelerating force, f
- **Dependent** (to measure): acceleration of trolley, a

➤

General method

1 Set up the ramp and trolley as shown in Figure 5.10.
   If the pulley does not have a suitable clamp attached to it, fix it firmly in place with a lab stand and clamp.

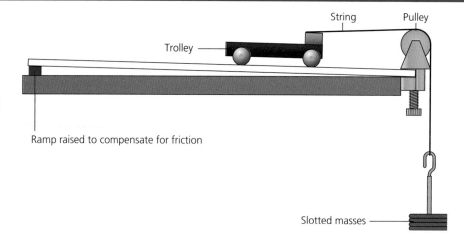

■ **Figure 5.10** Experiment setup for plane and trolley

2 Begin by piling all the masses on the trolley, so that only the mass hanger is 'pulling' the trolley.
3 Release the trolley to roll along the plane, and measure the acceleration using one of the three methods given below.
4 Now take a mass from the trolley and transfer it to the mass hanger. Release again and measure acceleration.
5 Repeat the experiment until all the masses have been used.

Hypothesis

Write a hypothesis about the motion of the trolley as the mass is transferred to the hanger.

Explain your prediction with reference to Newton's second law.

Measuring the acceleration

There are three different ways to measure the acceleration described on the following pages, depending on the equipment available.

After reading the method and analysis for your chosen acceleration technique, design a data table to record your 'raw' or unprocessed data from the experiment.

Hint

What is providing the accelerating force on the trolley? What is the total mass of the whole system that is being accelerated?

Method 1: using the ticker-tape timer

A ticker-tape timer has a small needle that vibrates at a known rate. The needle makes a mark on a long, thin piece of paper tape that runs through the timer. This gives a 'trail' of marks whose distance apart can be used to measure velocity, and so acceleration for the trolley.

To use the timer, fix one end of the tape to the trolley and pass the other end through the timer machine. When you are ready to release the trolley, turn on the ticker-tape timer first, then go!

Data analysis using method 1

You can use the tapes to produce a distance–time graph. Simply cut the tape for each 'run' of a different mass every 5–10 dots or so. Place them side by side in order (Figure 5.11).

You can then measure from the tapes to derive the initial and final velocity of your trolley, and so the acceleration of the trolley as shown on the diagram.

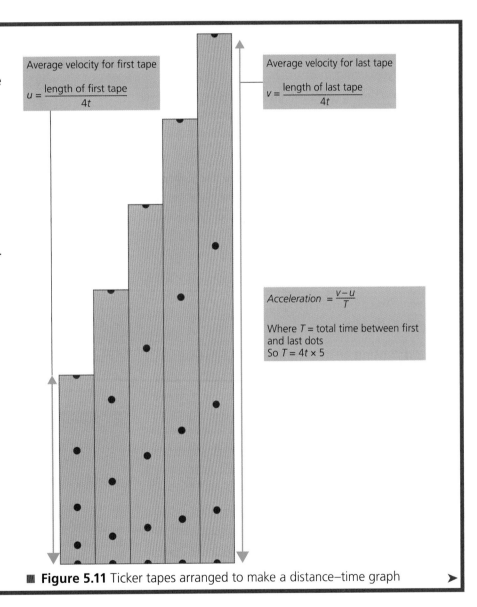

Average velocity for first tape

$$u = \frac{\text{length of first tape}}{4t}$$

Average velocity for last tape

$$v = \frac{\text{length of last tape}}{4t}$$

$$Acceleration = \frac{v-u}{T}$$

Where T = total time between first and last dots
So $T = 4t \times 5$

■ **Figure 5.11** Ticker tapes arranged to make a distance–time graph

Method 2: using photo-optical timing gates

Photo-optical timing gates send an electrical signal whenever the light source of the emitter is interrupted and light does not reach the receiver. To measure average velocity, you would need two timers – one at the beginning of the ramp and the other at the end.

To measure acceleration, you will need four timers – two to measure the velocity at the beginning of the ramp, and two to measure the velocity at the end of the ramp.

You should set the timers so that they run continuously but 'freeze' when the trolley passes by. This gives you the time taken, t, by the trolley to pass each gate.

Data analysis using method 2

The first two timing gates will give you the time of travel at the beginning of the ramp. Measure the distance, d, between the timers and divide by this time for the initial velocity, u.

The second two timing gates will give you the time of travel at the end of the ramp. Measure the distance, d, between the timers and divide by this time for the initial velocity, v.

Calculate acceleration using $(v – u)$ divided by the total time between the first gate and the last gate, that is,

'frozen' time shown on last gate – 'frozen' time shown on first gate

Method 3: using an ultrasonic distance measurer

An ultrasonic distance measurer emits an audible 'click' which is then reflected back from an object and picked up at the receiver (which is usually just a directional microphone built into the unit). The travel time for the 'click' and its reflection is used to measure the distance to an object (see Chapter 6 for more on sonar detection). You may need to use a square of card attached to the trolley to reflect the ultrasonic click effectively.

The distance and time are usually displayed on a datalogger, or computer software, both as a data table and graphically.

Data analysis using method 3

The output from the ultrasonic measurer will give you a constant stream of distances over the whole time of the journey. From these, you can select points to calculate

initial and final velocities, then calculate acceleration as for the ticker-tape timer above.

Interpreting data

- **Design** a second data table to present your calculated or 'processed' values.
- What should you plot on a graph in order to get a straight line from your data?
- **Present** your processed data on a graph showing how the independent variable (force on trolley) affected the dependent variable (acceleration of trolley).
- Use your graph to determine a value for the mass of the whole system, M.
- From the value of M you have found, use the total mass of the hanger + 100 g masses to **deduce** a value for the mass of the trolley, m.

Conclusion

Compare the motion demonstrated in your results to the motion you predicted earlier. Was your prediction correct?

Now use a balance to find the actual mass of the trolley. Compare the value of m you obtained experimentally to your measured value. Were you correct?

Evaluation

- What problems (sources of error) were there with your experiment?
- **Calculate** the percentage error in the experiment by comparing the experimentally measured trolley mass, m, to the weighed mass of the trolley.
- What evidence was there in the data that these sources of error were significant?

> **Hint**
>
> How straight was your line? Did the line pass through the origin? What other force(s) might be changing as you add mass to the trolley?

- How important (significant) were each of these problems?
- How could you modify the design of your experiment to remove or lessen these problems?

◆ **Assessment opportunities**

This activity can be assessed using Criterion C: Processing and evaluating.

How fast, how much faster do objects fall?

■ **Figure 5.12**
Felix Baumgartner shortly before his record-breaking jump

NEWTON'S SECOND LAW AND FREEFALL

Is it possible to parachute from outside the Earth's atmosphere? The answer is yes, and it has been attempted three times! (However, there isn't much point in using a parachute until you are well within the atmosphere.)

The first attempt to parachute from space was in 1960 for NASA by the American test pilot Joe Kittinger (find out about him and watch videos using **Kittinger jump**). Kittinger jumped from 31 km, at the outer edge of what we consider to be the atmosphere, or 'inner space' as it was sometimes called by astronauts. More recently the stunt was repeated by the Austrian Felix Baumgartner, who used a balloon to lift him in a pressurized capsule to an altitude of 39 km and then jumped!

ACTIVITY: Freefalling motion

■ ATL

- **Critical-thinking skills**: Gather and organize relevant information to formulate an argument; Evaluate evidence and arguments

Aim: To present and analyse the motion of a parachute jumper.

Use **http://www.youtube.com/watch?v=VKojXTWJIhg** to watch the view as Baumgartner jumps to Earth from inner space! The film also shows variables for his motion and position in the form of dials and readouts in the camera view.

Hypothesis

Outline what you think will happen to Baumgartner's speed and acceleration as he falls. **Explain** why you think this will happen.

Method

Use the film to gather data about the motion of Baumgartner during the jump. Choose a suitable interval for stopping the film and reading off your chosen data.

Variables

Independent: time (s). You can read the time from the bottom of the camera view in the top left image.

> **Hint**
> Not all variables may be listed in the display.

What other variables can you read from the display? Which variables are needed for your analysis? Which variables do you think they attempted to keep constant?

Data presentation and analysis

Show your data on suitable graphs.

Identify the points where Baumgartner's

- **speed was at a maximum**
- **acceleration was at a maximum**
- **speed was constant**
- **acceleration was negative.**

Conclusion

Were the results what you expected? Compare them to your prediction. **Outline** what is happening at different stages of Baumgartner's journey to Earth. **Summarize** the forces acting on Baumgartner at each stage of his fall.

◆ Assessment opportunities

In this activity you have practised skills that are assessed using Criterion C: Processing and evaluating.

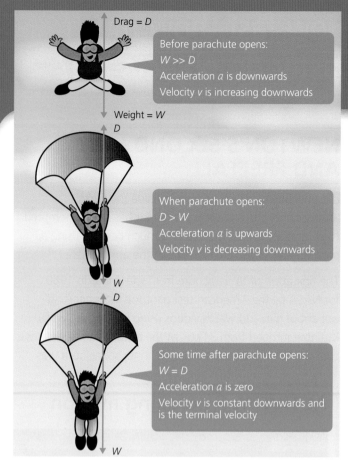

Before parachute opens:
$W \gg D$
Acceleration a is downwards
Velocity v is increasing downwards

Drag = D

Weight = W

When parachute opens:
$D > W$
Acceleration a is upwards
Velocity v is decreasing downwards

Some time after parachute opens:
$W = D$
Acceleration a is zero
Velocity v is constant downwards and is the terminal velocity

■ **Figure 5.15** Weight and drag forces on a parachutist

When objects fall through a fluid, the gas (or other fluid) exerts a retarding force called **drag** on the object – something you may have experienced if you have ever cycled on a windy day!

If the object is falling through a vacuum, there is only one force to accelerate it: **weight**.

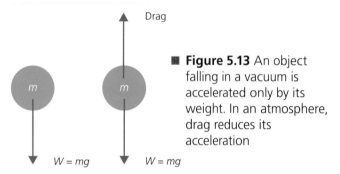

Drag

$W = mg$ $W = mg$

■ **Figure 5.13** An object falling in a vacuum is accelerated only by its weight. In an atmosphere, drag reduces its acceleration

Since weight is proportional to mass (see Chapter 2),

$w = mg$

a greater mass has a proportionately greater force (weight) produced by the gravitational field on it. This means that **no matter what mass objects have, acceleration in a gravitational field is always the same**.

This seems counter-intuitive. It's true that 'light' objects – such as feathers – fall more slowly than 'heavy' objects – such as hammers. Aristotle thought this was because of the elements they contained, and so the rate of fall for him was related to

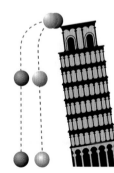

■ **Figure 5.14** Galileo's demonstration of freefalling masses. Galileo used two identically sized balls, one solid iron and the other made from wood. After simultaneous release from the top of the tower, the two balls always hit the ground at the same time

the matter in the object. However, the different rate of fall isn't because of their different masses, but because of the different drags produced on them. This was first realized by Galileo, who reputedly carried out an experiment at the tower of Pisa to demonstrate this (Figure 5.14).

To really prove the point, on the *Apollo 15* moon mission of 1971 astronaut David Scott actually dropped a feather and a hammer on the moon's surface, where there is no atmosphere.

Use: **Apollo 15 hammer and feather** to locate a video of this experiment.

While the size of the weight depends only on the mass of the object (which does not usually change), the drag depends on the velocity: as you will notice if you ever stick your hand out the window of an accelerating car. As the object falls and accelerates, the drag force increases (Figure 5.15).

At some point, the drag and the weight will be equal in size, but opposite in direction – so they **balance**. The resultant force on the object is then zero and there is no further acceleration. The velocity at which this occurs is called **terminal velocity**.

ACTIVITY: Investigating aerodynamics and freefall

■ **Figure 5.16** A French Train à Grande Vitesse or TGV

When designing fast vehicles, such as jet aircraft and high-speed trains, engineers have to take drag seriously. At high velocities, drag can not only reduce efficiency and so increase fuel consumption, it can even seriously destabilize the vehicle.

In this investigation you will model the effects of drag using a fluid in a vertical tube. You will need to decide on the different objects to drop through the fluid, and work out how to compare them.

Basic equipment
- **Long tube, measuring cylinder or similar**
- **Thick fluid, such as olive oil**
- **Modelling putty**

You should select the appropriate equipment to make measurements in your investigation.

Use the investigation cycle below to design, carry out and conclude your investigation (Figure 5.17). In your design, **explain** why you need to use a thick fluid to investigate drag force, rather than simply using air.

Safety: Make sure you check your design with your teacher for safety before starting.

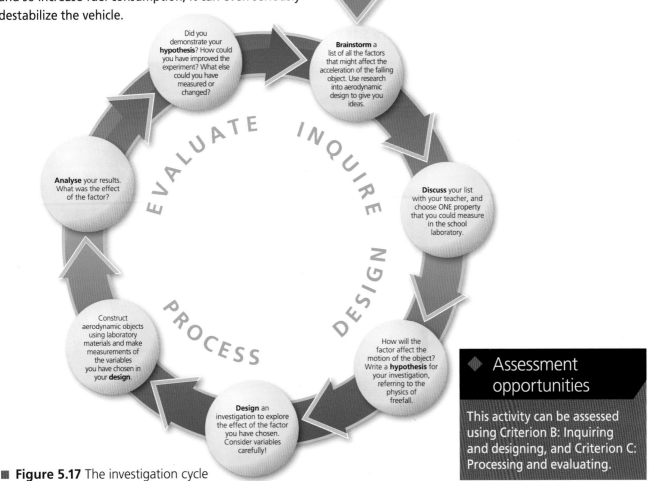

■ **Figure 5.17** The investigation cycle

THE THIRD LAW OF MOTION

Read Newton's third law of motion in Figure 5.9 (page 75).

The invention of rockets dates back at least as far as the discovery of the mixture of substances that gave us gunpowder. The discovery is usually credited to alchemists in China around 850 CE who were experimenting to find an immortality potion. By the 13th Century CE the Chinese were reported to be using 'fire arrows' made from bamboo stuffed with gunpowder.

The jet engine, on the other hand, took some time to develop – the first working prototype was built by Sir Frank Whittle in Derby, England, in 1937. The jet engine has made intercontinental air travel an affordable reality for many; the rocket, meanwhile, has been developed to the point where it has propelled us for the first time beyond the confines of our blue planet and into space.

■ **Figure 5.18** Woodcut print of early Chinese rockets

■ **Figure 5.19** Sir Frank Whittle's first working jet engine

ACTIVITY: Distinguishing good hypotheses from bad

What makes a rocket or a jet work? How do they produce **thrust** for aircraft and spacecraft?

Individually, write down your answer to the questions above.

Some MYP 4 students did a survey to find out what ideas students and adults had in their school and local community about jet or rocket propulsion. They asked 20 people the same question: 'Why does a rocket move?' They then collected the answers together and recorded the ones most people gave. Next, they took only the wrong answers and put them in order of popularity. The most common **wrong** answers were as follows:

1 **The rocket burns fuel and the hot gases that come out of the back push against the air to make the rocket move upwards.**

2 **The gases that come out of the bottom of the rocket push against the ground and make the rocket go upwards.**

3 **The gas that comes out of the bottom of the rocket expands and the expansion lifts the rocket upwards.**

4 **The gas that comes out of the bottom of the rocket is hot and hot gases rise, so the rocket rises.**

Work in pairs

Discuss each of the wrong explanations. **Outline** an argument that would prove each explanation to be wrong. **Suggest** one experiment you could do to demonstrate that each explanation was wrong.

Proving hypotheses wrong is often more important in science than gathering evidence to support hypotheses. After all, how much evidence do you need to prove something is **always** correct? How many times do you need to prove a hypothesis is wrong? This idea in science is called **falsification** and it was developed at length by the philosopher Sir Karl Popper (1902–1994).

So, what is the right answer?

In Newton's third law of motion, the **action** referred to is a force acting between two objects. Newton's law states that whenever a force **acts**, then there is another force acting back again that is equal in size and opposite in direction.

Newton conceived this law when working on the idea of gravitation. He realized that objects in orbit must exert equal and opposite forces on each other – more on this in Chapter 8, when we look at orbits in detail. One consequence of the law, however, is that **the total momentum in a closed system must remain constant**. The 'system' is closed as long as no force acts from outside.

We can consider a loaded cannon to be a closed system. Before the cannon is fired, the cannon and the cannonball are stationary, so velocity is zero.

The balloon rocket (Figure 5.22) works on the same principle. No bullet is being fired, but the gas in the inflated balloon and the 'rocket' all have zero momentum at first. When the gases are released rapidly from the balloon, they gain momentum backwards. This produces a reaction on the rocket itself, which is the **thrust**.

The principle is the same for jet engines and spacecraft – the only difference is the amount of thrust produced and the time it is produced for!

$p_{ball} = 0$ $p_{cannon} = 0$

■ **Figure 5.20** A cannon ball fired from a stationary cannon is a closed system

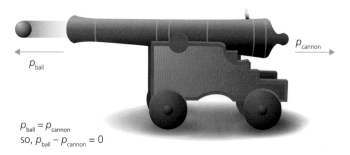

p_{ball} p_{cannon}

$p_{ball} = p_{cannon}$
so, $p_{ball} - p_{cannon} = 0$

■ **Figure 5.21** Momentum is conserved before and after the cannon is fired

After the cannon is fired, the cannonball gains momentum from the explosion of the gunpowder. However, the **law of conservation of momentum** tells us that we can't simply generate momentum from nowhere. There has to be an equal and opposite **reaction** that cancels out the momentum of the cannonball. This reaction appears as momentum of the cannon backwards. The cannon recoils when it fires the cannonball.

Since the mass of the cannonball is much smaller than the mass of the cannon, the effect of the momentum on the cannon is to move it more slowly than the cannonball. But the total momentum in the system is still zero after firing.

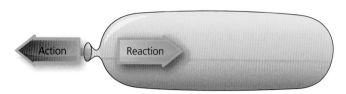

Action   Reaction

■ **Figure 5.22** A balloon rocket

ACTIVITY: Jet-propelled trolleys

■ ATL

■ **Creative-thinking skills:** Design new machines, media and technologies

Work in pairs.

Your task is to use the equipment below to make a jet-propelled dynamics trolley that will move under its own propulsion for 10 cm.

Equipment
- **Balloons of different shapes and sizes**
- **A dynamics trolley**
- **Some sticky tape**

(You may use additional materials as well.)

Evaluation
How well did your jet trolley work? **Outline** the forces acting on the trolley as the balloon deflates. **Suggest** improvements to the design to make it work better.

ACTIVITY: Is sooner better than later?

Many countries have seen high-speed railway as one answer to their growing transportation needs. But not everybody agrees that this is the solution.

Table 5.1 gives a comparison of high-speed railways worldwide.

Research to find out about controversies surrounding some major high-speed rail projects around the world.

Find out:

- **How a high-speed railway works.**
- **What scientific and technological challenges are there in making it work?**
- **What do supporters of these transportation systems say are the advantages of such railway systems? Consider factors such as the socioeconomic, political, moral and environmental impacts.**
- **What do critics of these transportation systems say are the disadvantages of such railway systems? Consider factors such as the socioeconomic, political, moral and environmental impacts.**

Write a newspaper article, or a script for a radio or TV documentary about one of the high-speed rail projects you have researched.

■ **Table 5.1** High-speed lines by country (km). Source: UIC International Railway Union (2014).

Country	In operation	Under construction	Planned	Total
France	1896	210	2616	4722
Germany	1285	378	670	2333
Italy	923	0	395	1318
The Netherlands	120	0	0	120
Portugal	0	0	1006	1006
Russia	0	0	650	650
Spain	2056	1767	1702	5525
Sweden	0	0	750	750
Switzerland	35	72	0	107
United Kingdom	113	0	204	317
China	6299	4339	2901	13539
India	0	0	495	495
Japan	2664	378	583	3625
Saudi Arabia	0	550	0	550
South Korea	412	186	49	647
Turkey	447	758	1219	2424
Brazil	0	0	511	511
USA	362	0	900	1262
Total world	17166	8838	16318	4232

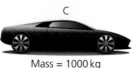

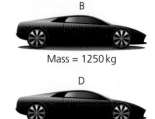

A	B
Mass = 1500 kg	Mass = 1250 kg
C	D
Mass = 1000 kg	Mass = 1750 kg

■ Figure 5.24

SOME SUMMATIVE PROBLEMS TO TRY

Use these problems to apply and extend your learning in this chapter. The problems are designed so that you can evaluate your learning at different levels of achievement in Criterion A: Knowledge and understanding.

THIS PROBLEM CAN BE USED TO EVALUATE YOUR LEARNING IN CRITERION A TO LEVEL 1–2

1 **State** which of these statements is true for an astronaut in a space shuttle in orbit around the Earth, then **justify** your answer.

　A　Her mass and her weight are zero.

　B　Her mass and her weight are the same as on Earth.

　C　Her mass is the same as on Earth, her weight is zero.

　D　Her mass is zero, her weight is the same as on Earth.

2 **State** which force arrow on the diagram below best represents the force produced by the jet engines.

■ Figure 5.23

3 If the cars in Figure 5.24 all have an engine thrust of 1000 N, **state** which one will have the greatest acceleration. **Justify** your answer with reference to Newton's laws.

THIS PROBLEM CAN BE USED TO EVALUATE YOUR LEARNING IN CRITERION A TO LEVEL 3–4

4 Kasper is sitting in a boat in the middle of a lake.

　In the boat he has 10 fish, each of mass 2 kg. The boat has mass 100 kg, and Kasper's mass is 40 kg. Unfortunately, he falls asleep, and loses the oars of the boat.

　Then he has an idea. He decides to make the boat move by throwing the fish out of the boat.

■ Figure 5.25

　a If he throws a fish out of the rear of the boat, **state** which way the boat might move. **Explain** why this happens.

　b If he throws a fish with velocity 1.5 m s^{-1}, **calculate** the momentum the fish will have.

　c After he throws the fish, **state** the momentum that the boat and Kasper will have. **Justify** your answer.

　d **Calculate** the velocity that the boat and Kasper will have just after throwing the fish.

　e If Kasper throws one fish every second, **describe** how the velocity of the boat will change after he has thrown out all 10 fish. (Assume the resistance from the water is negligible.)

　f Is this method likely to get Kasper to the shore and out of trouble? **Outline** the problems of moving the boat this way.

THIS PROBLEM CAN BE USED TO EVALUATE YOUR LEARNING IN CRITERION A TO LEVEL 5–6

5 a A rifle bullet is fired with a velocity of $240\,\text{m}\,\text{s}^{-1}$ and a mass of 100g. A van has a velocity of $1\,\text{m}\,\text{s}^{-1}$ and a mass of 2.4 tonnes. **Compare** the momentum of both.

   b **State** the size of the momentum of the rifle after the bullet is fired.

   c The bullet hits a target and is stopped in 10ms. **Outline** the forces acting when this occurs and **calculate** their size and direction.

   d The van accelerates from $1\,\text{m}\,\text{s}^{-1}$ to $5\,\text{m}\,\text{s}^{-1}$ in 10s. **Calculate** the extra force produced by the engine during this time.

   e A strong wind starts to blow and the van's speed drops by 10% in 10s. **Calculate** the additional drag produced by the wind.

   f The van turns around a corner, still with constant speed of $5\,\text{m}\,\text{s}^{-1}$. Your teacher tells you that the van is accelerating. **State** whether you agree or not. **Explain** your reasons.

THIS PROBLEM CAN BE USED TO EVALUATE YOUR LEARNING IN CRITERION A TO LEVEL 7–8

6 A spherical submarine controls its depth underwater by varying its **buoyancy**. The buoyancy produces an upward force on the submarine's hull called the **upthrust**, which acts in opposition to the submarine's **weight**. When the submarine moves up and down through the water, there is also a force of **drag** that is proportional to the density of the water and to the velocity of the submarine.

   a On suitable diagrams similar to those below in Figure 5.26, draw force arrows to **outline** the forces acting in each of the following situations. **Describe** the relative sizes and directions of the forces acting.

   ■ The submarine is floating stationary under water.
   ■ The submarine is sinking at a constant rate.
   ■ The submarine is accelerating upwards to the surface.

   b A submarine has a mass of 5 tonnes. **State** the buoyancy required for it to float stationary in the water.

   c The submarine expels some air to reduce its buoyancy by 10%. It now has 'negative buoyancy.' **Calculate**

   ■ the size of the resultant force on the submarine, and **state** its direction
   ■ the acceleration of the submarine immediately after reducing buoyancy.

■ **Figure 5.26**

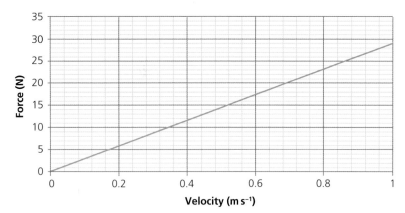

Figure 5.27 Drag with velocity for a 1.5 m radius spherical submersible

Stokes' law relates the drag force, F_d, on a spherical object to its velocity, v, its radius, R, and the density of the fluid through which it is passing, ρ:

$$F_d = 6\pi\rho Rv$$

Figure 5.27 shows how the drag increases with velocity for our submarine.

d With reference to the graph, **outline** how the drag changes as the submarine accelerates through the water.

e **Summarize** the motion of the submarine after it decreases its buoyancy.

As the submarine sinks ever deeper into the ocean, the density of the seawater changes as shown in Figure 5.28.

f As the submarine descends, **describe** how the density of the seawater changes.

g With reference to Stokes' law and to the information about the density of seawater above, **summarize** the motion of the submersible as it descends.

h **State** and **explain** what action the divers must take if they are to maintain a constant velocity of descent.

i **Evaluate** during which part of the dive this action will be most important.

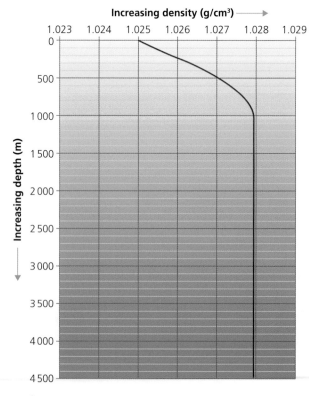

Figure 5.28

Reflection

In this chapter we have explored the ways in which force causes change in motion, while matter prefers to carry on doing what it was doing before – **inertia**. We have considered how momentum can be used to quantify the motion of an object, and finally we have considered how Newton's laws of motion help us understand and develop new forms of propulsion.

Use this table to reflect on your own learning in this chapter		
Questions we asked	Answers we found	Any further questions now?
Factual: What causes motion? What changes motion?		
Conceptual: What relationship is there between force and motion?		
Debatable: Does everybody have the right to travel the way they wish? Should everybody have the right to travel as they wish?		

Approaches to learning you used in this chapter	Description – what new skills did you learn?	How well did you master the skills?			
		Novice	Learner	Practitioner	Expert
Communication skills					
Critical-thinking skills					
Collaboration skills					
Creative-thinking skills					
Transfer skills					
Research skills					
Learner profile attribute(s)	Reflect on the importance of being open-minded for our learning in this chapter.				
Open-minded					

6 How do we make life easier?

Human life has been completely **changed** and **developed** through the use of machines that are created to **transform energy** and do useful work.

Figure 6.1 Woman cooking over a traditional oven

CONSIDER AND ANSWER THESE QUESTIONS:

Factual: What are examples of energy? What is temperature? How can we measure the work done? How does heat energy transfer? How can we measure the efficiency of machines?

Conceptual: What is energy? How is energy harnessed to do work? How does energy interact with matter? How is energy transformation controlled? How much of the energy available can be used to do useful work?

Debatable: Do the machines we make always change things for the better? Will energy ever 'run out'?

Now **share and compare** your thoughts and ideas with your partner, or with the whole class.

IN THIS CHAPTER, WE WILL...

- **Find out** how energy is related to change as **work done**, how machines are designed to control the way that energy is transformed as work, but also how machines are wasteful.
- **Explore:**
  - the interaction of heat with matter
  - how heat is transferred and formulate what we actually measure when we measure temperature
  - the meaning of energetic efficiency and power
  - and evaluate the ways in which even simple machines can make a big difference to human lives.
- **Take action** to campaign for more equal access to those machines in a world where some people live surrounded by machines, while others still have to rely heavily on the resource of their own body.

Energy makes change happen. In order to survive, all living things need to use energy. One way to understand life is as a process that uses energy to transform environmental conditions so that they are better for life. This applies to humans just as it does to other kinds of life, and access to energy is a key issue for development globally.

- **Information literacy skills**
- **Communication skills**
- **Critical-thinking skills**
- **Creative-thinking skills**
- **Transfer skills**

● We will reflect on this learner profile attribute …

- Caring – we will have the opportunity to reflect on how even simple machines can make a big difference to the lives of those in less economically developed countries.

◆ Assessment opportunities in this chapter

- ◆ **Criterion A**: Knowing and understanding
- ◆ **Criterion B**: Inquiring and designing
- ◆ **Criterion C**: Processing and evaluating
- ◆ **Criterion D**: Reflecting on the impacts of science

KEY WORDS

efficiency	power
energy	transform
machine	stored
potential	work

ACTIVITY: Energy and development

■ ATL

- **Information literacy skills**: Access information to be informed and inform others

- **Making better stoves for cooking:** http://www.africancleanenergy.com/
- **Energy for opportunity in Sierra Leone:** http://youtu.be/WON3XvKjisw
- **European Union sustainable energy projects in Somalia:** Somali Energy and Livelihoods Project

In groups watch one of these online videos about the ways in which the science and technology of energy is being used to help people better their conditions of life in disadvantaged parts of the world. While watching the videos, make notes around these guiding questions:

- **What are the energy challenges of the places shown?**
- **What methods are shown for using energy more effectively?**
- **After watching your video, feed back to other groups about the case study.**

In this topic we will be exploring the physics of energy. We have already seen how force fields can make changes happen at a distance (Chapter 2) and at how forces cause motion (Chapter 5). We will now consider how energy is related to the concept of force in physics.

What are examples of energy?
What is energy?
What is heat?

EVERYTHING HAS ENERGY

When you hear the word 'energy', what do you think of? Perhaps the Sun, or the way you feel after a good breakfast at the start of the day, or the way a younger brother or sister or cousin can go on playing for hours on end without ever seeming to get tired!

In order to begin to get a handle on such a big concept, it is a good idea to list the things we consider may have energy, and to see whether they can be grouped or categorized according to the kinds of energy they have.

One starting point in categorizing energy types is to consider whether the energy is being stored, or whether it is actively doing something.

Some other categories that we can use to understand energy forms are:

- **Kinetic energy**: energy that causes displacement of an object, that is movement.
- **Mechanical energy**: energy held in the physical bonds that hold a material together.
- **Chemical energy**: energy held in the chemical bonds between atoms within molecules.
- **Thermal energy**: energy due to the motion of particles in an object or substance, even if the object itself is not moving; present whenever the temperature of a substance can be measured.
- **Electrical energy**: energy due to differences in electrical charge (positive or negative).
- **Magnetic energy**: energy due to the interaction of magnetic dipoles in a suitable material.
- **Nuclear energy**: energy held in the nucleus of the atom, between nucleons (protons and neutrons).

■ **Figure 6.2** Various energetic objects

ACTIVITY: Kinds of energy

■ ATL

■ **Critical-thinking**: Gather and organize relevant information to formulate an argument

Look at the selection of objects listed below, and consider these questions:

- **Do they have any energy?**
- **Where is their energy?**
- **How much do they have?**
- **Is the energy 'active' or 'stored'?**
- **What kind of energy do you think they have?**
- **Do they have more than one kind of energy?**

Think-pair-share and then **organize** your ideas in the form of a diagram or table.

- **A piece of pasta**
- **Beaker of hot water**
- **Ice cube**
- **Stretched elastic band**
- **Compressed spring**
- **A bar magnet**
- **1.5 V battery**
- **A water wave**
- **A doorbell or buzzer**

Share your ideas with other groups.

As a class, try to **discuss and agree** on a definition that begins: 'Energy is …'

We have already suggested that energy might be present, even if it is not in fact doing anything – this is called stored or **potential energy**.

Still another way to think of energy is by considering its **relationship** with force. Whenever a force is happening, then energy must be available to **change** something. Thinking back to Chapter 2, we saw that force is **always** around in all parts of the Universe – so energy must always be available too! So why does it seem to be so difficult to get energy to work for us?

■ **Figure 6.3** More energetic objects

How is energy harnessed to do work? How can we measure the work done?

PUTTING ENERGY TO WORK

Putting energy to work has been a concern for humanity from time immemorial – the more we can get machines to work for us, the more free time we have to do other things! Nevertheless, much of our current understanding of energy derives from a particular period in a particular place – a period referred to by historians, economists and others as the Industrial Revolution, which began in Northern England in the 18th and 19th centuries.

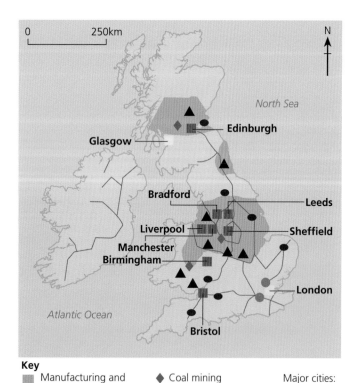

Key

- ▨ Manufacturing and industrial areas
- — Railways by 1850
- ● Banks
- ◆ Coal mining
- ● Iron industry
- ▲ Textile industries

Major cities:
- ▨ 1820
- ▨ 1850

■ **Figure 6.4** Map of the UK in the 19th century CE showing regions of the Industrial Revolution

▼ **Link:** The history of humanity as energy users

■ **ATL**

■ **Transfer skills:** Inquire in different contexts to gain a different perspective

You will have studied the history of different civilizations in different parts of the world. It enriches our scientific understanding to consider questions from a historical perspective, too:

- To what extent was the development of civilizations influenced by the kinds of energy they had available to use?
- What is the timeline for different energy uses for a particular civilization?

The Industrial Revolution was the result of a number of different factors, all converging in the same place at the same time. There was the discovery of fossil fuels – particularly coal – in abundance, along with iron ore and the technology to turn it into metal iron and steel. There was a large population available to use as workers in the new industries that would be powered by the new machines. There was a growing demand for manufactured goods, brought about by rising wealth from trade and exploitation of raw materials and other resources in other countries. Finally, there was the availability of capital – money – to invest in new projects, provided a good return as profit looked probable to the investor.

■ **Figure 6.5** The Industrial Revolution resulted in workers being concentrated in large mills or factories where living conditions were often very hard

Table 6.1 Energy available in some common objects and processes

Object or process	Approximate energy
To boil 1 l of water from room temperature	320 kJ
Stored in a 100 g ice cube at 0°C	6.3 kJ
In a piece of pasta (e.g. penne)	49 J
Stored in a 1.5 V battery	5.4 kJ

Perhaps the early pioneers of industrialization, such as Richard Arkwright, James Watt and Matthew Boulton, also had the requisite imagination, curiosity and passion to make things happen. However, it is clear that the drive to produce better and better machines to put energy to work was propelled by the need to return a good profit on investment.

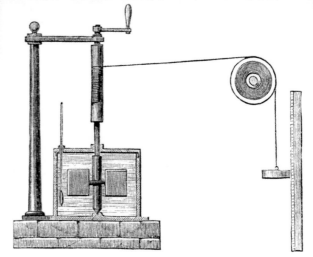

Figure 6.6 Joule's experiment for measuring work done by heat

James Joule (1818–1889) wanted to devise a way to measure accurately the energy available from heat using a machine. In doing so, he quantified the concept of energy and the unit of energy is named after him – the **Joule**.

ACTIVITY: How energetic is my food?

All mammals – including humans – gain their energy from the **aerobic respiration** of foods.

Nutrition				
Typical values	100 g contains	Each slice (typically 44 g) contains	% RI*	RI* for an average adult
Energy	985 kJ	435 kJ		8400 kJ
	235 kcal	105 kcal	5%	2000 kcal
Fat	1.5 g	0.7 g	1%	70 g
of which saturates	0.3 g	0.1 g	1%	20 g
Carbohydrate	45.5 g	20.0 g		
of which sugars	3.8 g	1.7 g	2%	90 g
Fibre	2.8 g	1.2 g		
Protein	7.7 g	3.4 g		
Salt	1.0 g	0.4 g	7%	6 g
This pack contains 16 servings				
*Reference intake of an average adult (8400 kJ/2000 kcal)				

Figure 6.7 Sample food package nutritional information box

Food energies are commonly expressed as **calories (cal)** or **kilocalories (kcal)** where

1 calorie = 4.18 Joules

All food packaging carries nutritional information which includes the 'calorific' or energy content for a fixed mass of food.

Analyse a meal during your day for its calorific content.

Make a **survey** of the calorific content of different daily meals in your class or **set up a blog** so that different members of the class can enter their daily calorific intake anonymously for your analysis.

Research: What is the typical daily energy requirement for a young adult? What factors affect this energy requirement the most?

Summarize your findings in a **blog** or other online social network so that other young adults can learn from your findings.

◆ Assessment opportunities

In this activity you have practised skills that can be assessed using Criterion A: Knowing and understanding.

How is energy harnessed to do work? How can we measure the work done?

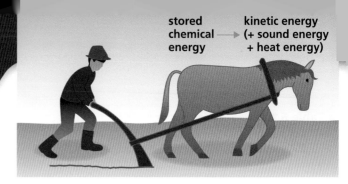

stored chemical energy → kinetic energy (+ sound energy + heat energy)

■ **Figure 6.8** Energy changes for a horse pulling a plough

To make change happen, we have to transform energy from one form to another. Throughout history, humans have designed **machines** to control the way energy is transformed, such that it does something useful along the way, and the amount of energy transformed is then measured as the **work done** by an energy changer.

One of the biggest changes in the Industrial Revolution was the replacement of animals as a source of energy with machines. In Western Europe, the chief source of energy for work for millennia had been the horse. We can represent the energy transformations when a horse does work using an energy flow diagram (Figure 6.8).

The inputs for the system are the sources of energy for the horse, such as the food the horse is given. The output is then the useful energy taken out of the system as work – in this case, the kinetic energy gained by the plough.

A horse could also be used to drive a mill to grind corn, but for this humans devised more **efficient** and reliable machines that used natural energy resources. At first, these were wind (in windmills) or water (in watermills).

Finally, a way was found to harness the heat energy released by the burning or **combusting** of fossil fuel, primarily coal. The **steam engine** uses the heat from combusting coal to boil water to make steam. The steam is contained at high pressure until the combined kinetic energy of the water molecules is enough to drive a piston and turn a wheel.

stored gravitational energy → kinetic energy (+ heat energy + sound energy)

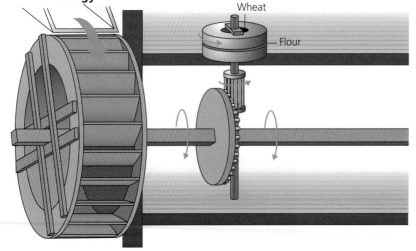

Wheat

Flour

■ **Figure 6.9** Energy changes for a water mill

ACTIVITY: Windmills, a past and future energy changer?

Look at Figure 6.10 which shows a windmill used for grinding corn, and a modern wind turbine used for generating electricity.

Research the use of machinery to 'harness' the energy of the wind in windmills and wind turbines.

Draw energy-change diagrams for each.

Evaluate the advantages and disadvantages of using wind as a source of energy.

■ **Figure 6.10** 'Windmills', past and present

◆ Assessment opportunities

In this activity you have practised skills that can be assessed using Criterion A: Knowing and understanding.

ACTIVITY: Full steam ahead!

Research how steam engines work, and what they were used for. Watch some animations of steam engines in action using **steam engine animation**. **Find out** what the following parts were, and what they did:

piston   boiler   crank   valves.

Draw an energy-change diagram for a steam engine.

Evaluate the advantages and disadvantages of using steam engines as a source of energy.

◆ Assessment opportunities

In this activity you have practised skills that can be assessed using Criterion A: Knowing and understanding.

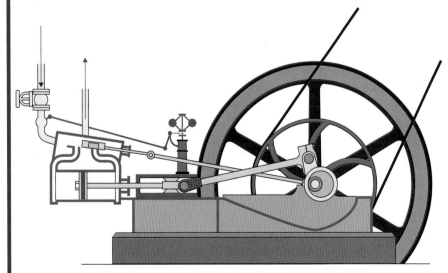

■ **Figure 6.11** Schematic diagram of a steam engine

ACTIVITY: Using gravity as an energy source

ATL

- **Information literacy skills**: Collect, record and verify data; Process data and report results
- **Creative-thinking skills**: Design improvements to existing machines, media and technologies

Aim: To measure the work done by gravity acting on a weight.

Background

From Chapter 2, the gravitational force accelerating a mass is the **weight**, w

$$w = mg$$

Therefore, the **work done** by a mass falling from a height, h, will be

$$\text{work done} = Fd = mgh$$

This is also the **gravitational potential energy** in the mass before it is released.

The kinetic energy due to motion of a mass at velocity, v, is given by

$$KE\ (\text{J}) = \frac{1}{2}mv^2$$

In this experiment you will measure the work done by a machine that uses gravitational energy as its energy input.

Equipment

- **Mass hanger and 5 × 100 g masses**
- **String**
- **Bench pulley**
- **Ticker-tape timer, or ultrasonic distance measurer**

Variables

- **Independent** (to control): accelerating mass, m; height, h, of drop
- **Dependent** (to measure): constant final velocity v of trolley

Method

Arrange the apparatus as shown in Figure 5.10 (page 76). Note this is the same setup as was used to work out the acceleration of a trolley!

As before, allow the masses to fall and so accelerate the trolley. However, adjust the height, h, through which the masses fall such that the masses stop falling before the trolley reaches the end of the ramp. Leave the ticker-tape or other velocity-measurement device running for this time.

  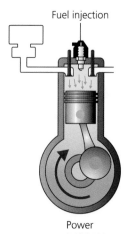 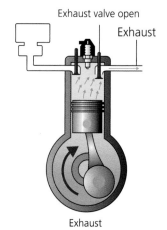

Intake valve open — Air — Intake

Both valves closed — Compression

Fuel injection — Power

Exhaust valve open — Exhaust — Exhaust

■ **Figure 6.12** Piston of an internal combustion engine

You can chose either to vary the mass, m, or the heigh, h, of the drop, and carry out repeat 'runs'. (Remember: only change **one variable at a time!**)

Analysis

After the masses have stopped pulling the trolley, the trolley will roll with constant velocity, v, since there is no longer an accelerating force. You should be able to identify this time on the ticker tapes, or in the output from the distance measurer.

Find the velocity, v, of the trolley gained after acceleration has stopped.

Organize and **present** your data to show how v changed with your chosen independent variable.

Conclusion

Using the equations above and your data, **calculate** how much of the gravitational potential energy of the masses was transferred to kinetic energy in the trolley.

Evaluation

Were your values for gravitational potential energy, PE, and for kinetic energy, KE, the same?

What has happened?

Compare your expected values for PE and the measured values of KE as a percentage error in the experiment.

Identify the likely sources of this error.

Discuss your evaluation with another group. Consider these questions:

- **Was it true that the trolley moved with constant velocity, v, after the masses stopped falling?**
- **What other forces might be acting on the trolley during this time?**

A useful technique: compensating for friction

In experiments involving dynamics trolleys, we can compensate for the force of friction quite easily. Since friction stops the trolley from rolling freely, tip the ramp slightly until the trolley **just** begins to move.

This means that the force accelerating the trolley down the ramp must now be approximately equal to the force of friction stopping the trolley from moving.

Repeat your experiment with friction compensated.

How significant a difference has this made to the results? **Calculate** the new percentage error in your experiment and **compare**.

In your evaluation **outline** how friction could be further reduced as a source of energy loss in your machine.

 Assessment opportunities

This activity can be assessed using Criterion C: Processing and evaluating.

Of course, we have mostly moved on from steam engines as a source of work – although not entirely perhaps (see Chapter 10). The first replacement for the steam engine was the internal combustion engine, which uses either petroleum (gasoline) or diesel as fuel. Combustion engines do not use the heat released from combusting fuel to produce steam, but the principle remains the same: the work is transferred using the expansion of hot gases made when the fuel explodes inside the piston itself.

The **work done** by a machine is equal to the energy changed by the machine in order to do something useful, so its unit is the **joule**. If the energy change involves kinetic energy, and the object is moved a distance d by a force F, then the work done is given by

$$\text{work done (J)} = F \text{ (N) } d \text{ (m)}$$

In the experiment above, the kinetic energy gained by the trolley is not equal to the gravitational potential energy lost by the falling masses because some energy is lost, particularly due to frictional forces. **Friction** is a serious problem in machines – to have an idea how serious, try rubbing your finger quickly along a table top. Press harder each time. You will notice that the friction increases rapidly, but also your finger will begin to feel hot. This is because friction has the effect of turning some of the energy in the system into **heat**.

How does energy interact with matter? How does heat energy transfer?

HEAT MATTERS

Although we do not regard everything as 'hot', everything – as far as we know – contains energy that can be transferred as heat. We tend to refer the 'hotness' of things to how they feel, which depends on the temperature of our own bodies – on average, 37°C. The amount of energy that is transferred between objects at different temperatures is often given the symbol Q, while temperature is given the symbol T.

- Heat (Q) is the total energy transferred between two objects at different temperatures, in the form of kinetic energy of particles in the objects.
- Temperature (T) is a measure of the average kinetic energy of the particles in an object.

A white hot metal 'spark' has a temperature near to 2000°C. A kettle of boiling water has a temperature of 100°C. Which of the two would be most dangerous in contact with your hand?

Although the spark is 'white hot', its mass is tiny, perhaps 0.01 g at most. So, it contains relatively little heat energy and might only feel like a pin-prick on your hand. A 1 litre kettle of boiling water, however, contains enough energy to seriously damage you and boiling water should never be poured over skin. The amount of heat is a total for all the particles, and the number of particles depends on the mass of the object. The temperature, however, does not depend on the mass, because it is an average over all particles (see Figure 6.14).

■ **Figure 6.13** Comparison of temperature and heat of 1 kg of boiling water and a sparkler

Theoretically, there is a temperature at which none of the particles are moving: here, the substance would have zero heat energy to transfer. This temperature is called **absolute zero**. The Celsius temperature scale is designed with reference to the freezing point (0°C) and boiling point (100°C) of water. Physicists sometimes use the absolute or Kelvin temperature scale instead, where

$$0\,K = -273°C$$

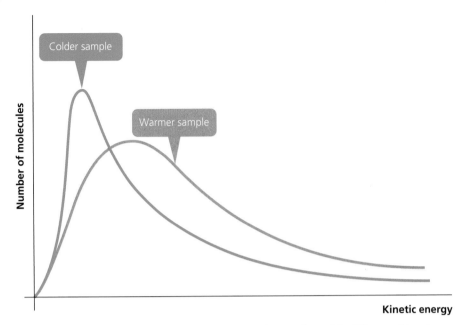

■ Figure 6.14 Graph showing the number of particles with different kinetic energies in a given mass. Note that because temperature measures the average kinetic energy in the mass, some particles will have more than this energy and some will have less. The total heat is the sum of all the energies under the curve.

It may seem strange to talk about the kinetic energy of the particles in a solid, when we are used to thinking of solids as an immobile mass. Yet, even within solids, the particles have a degree of freedom of movement. The amount of freedom a particle has to move (and so the amount of kinetic energy it can hold) depends on the **state of matter**.

If enough heat energy is provided or taken away, matter will change state as the kinetic energies of the particles change. Changing state means that bonds between particles are broken or formed, and this releases or absorbs energy.

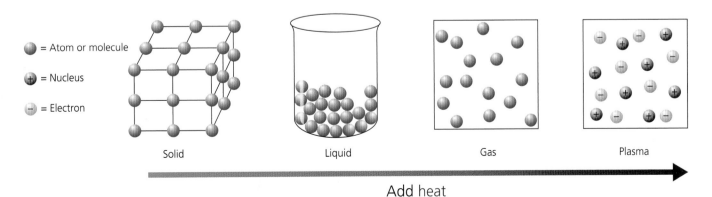

■ Figure 6.15 Particle motion in the states of matter

ACTIVITY: How cool is my ice?

Aim: To measure the energy required to melt an ice cube.

Background

The bonds between the particles of solid water in ice require energy to break them when the ice melts. When we mix ice with liquid water, the mixture tends towards a state in which the temperature is the same throughout, since heat from the liquid water is transferred into the ice. The solid bonds in the ice are broken as heat energy is taken from the water. The temperature of the ice increases and that of the water decreases towards new temperature for the ice–water mixture.

The heat energy, Q, required to change the temperature of a mass, m, by amount $(T_2 - T_1)$ is given by

$$Q = mc(T_2 - T_1)$$

where c is a constant for the substance called the **specific heat capacity**. The specific heat capacity for water is

$$c = 4200 \, \text{J kg}^{-1}\,°\text{C}^{-1}$$

The heat energy required to break the bonds and change state for a substance is given by

$$Q = mL$$

where L is a constant for a given material and denotes a state change called the **latent heat**. To melt ice to water the **latent heat** required is

$$L = 334 \, \text{kJ kg}^{-1}$$

Equipment

- One small ice cube
- A polystyrene drinking cup
- Thermometer, or digital temperature sensor
- Measuring cylinder
- Water at room temperature

Method

1 Fill the cup three-quarters full with water.
2 Pour this water into the measuring cylinder and measure the volume of the water.
3 Pour all the water back into the cup.
4 Place the thermometer or temperature sensor in the freezer compartment where the ice is being kept, or in the ice itself. Measure the temperature.
5 Place the thermometer or temperature sensor in the cup and measure the temperature of the water.
6 Take the ice cube from the freezer and quickly weigh it on a balance.
7 Drop the ice cube in the cup of water.
8 Measure the temperature of the mixture periodically, until all the ice has just melted into the water.

Analysis

Present your data in a form that allows you to judge when the temperature of the mixture has reached its minimum value. Record this value.

Conclusion

We will assume:

all energy into ice cube = energy out of the water

Calculate the mass, m, of the water before adding the ice. Remember that 1 ml of water at room temperature has a mass of 1 g.

> **Hint**
>
> There are **three** stages of heat exchange here:
> 1 The ice has to warm up to melting point.
> 2 The ice has to melt.
> 3 The mixture has to reach the final (minimum) temperature.

Use the equations for heat exchange in the background to **formulate** an equation for all three of these exchanges.

Using your measured values for temperature before mixing and after mixing, **calculate** the total heat, Q, that has been used to melt the ice cube.

What is the sign of Q? What is the significance of this?

Evaluation

What assumptions did we make in our calculations above? Are these assumptions valid?

What could we do to make our experiment more reliable, so that the assumptions we made were more likely to hold true?

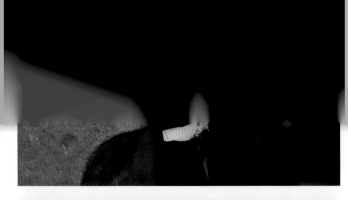

■ **Figure 6.16** Metals are very good conductors of heat; other substances (such as the ceramic tiles used to protect spacecraft on re-entry) are not

In the 'Activity: How cool is my ice?' heat is transferred from water to ice through direct contact of the two materials. This is an example of **conduction**. When particles are allowed to contact each other, energy is transmitted between them and so the available heat moves through the material. Some materials are better conductors than others – it depends on their molecular or atomic structure. Particularly bad conductors, such as the polystyrene cup in the experiment, are called **insulators**.

Metals are especially good conductors because they can transfer energy both through the bonds between atoms and also via the 'free' electrons which float around those atoms – like an energetic super-highway.

In liquids and gases – collectively referred to as **fluids** – it is more difficult for heat energy to move through direct contact of particles, since the particles are further apart and not strongly bonded. However, the increased kinetic energy of the particles at a higher temperature causes the particles themselves to move around more within the material, and so transfer kinetic energy through it. In regions of the fluid with more heat energy, the particles tend to be moving more quickly and the fluid is less **dense** than elsewhere, and so heat energy tends to spread to regions with less heat and greater density.

This process is known as **convection**. It explains why a small heater in one part of the room can heat the entire room quite quickly, since the rising air above the heater is replaced by denser, cooler air, which in turn is heated and rises, and so on to produce a circular convection current.

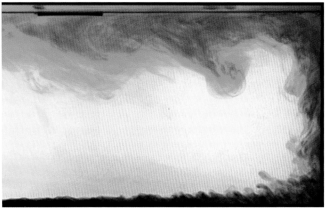

■ **Figure 6.17** Convection taking place in water, which is carrying a coloured dye with it. The right half of the bottom of the container is heated and the left half is cooled

■ **Figure 6.18** Earth's weather systems are, in large part, caused by the movement of heat through the atmosphere

There is one more way in which heat can get around. This is nothing to do with the motion of matter at all. Heat transfer by **radiation** occurs because heat energy can travel as photons – which we will explore in greater detail in Chapters 7 and 8. When travelling in this form, heat behaves rather like light – it tends to be absorbed by darker colours, and is reflected by lighter colours or shiny surfaces.

In order to keep a substance at a fixed temperature, we have to limit the heat transfer by all three of these processes.

■ **Figure 6.19** Marathon runners wrapped in silver foil to reflect their radiated body heat and prevent chill

ACTIVITY: The perfect container

Look at the diagram of the flask used to transport liquid medications (Figure 6.20).

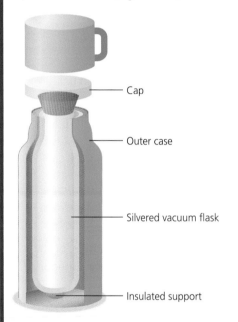

Cap

Outer case

Silvered vacuum flask

Insulated support

■ **Figure 6.20** Dewar or Thermos flask

Complete the table to show how each of the components of the flask reduces heat transfer.

Part of flask	Heat-transfer process
lid	
vacuum cavity (no air)	
silvery insides	
plastic casing	

If cost were not an issue, what could be done to improve the design of the flask? **Suggest** modifications to the design and **explain** how they would reduce heat transfer.

ACTIVITY: A better way to cook

In less economically developed countries, more efficient heat transfer can affect people's chances of survival. The **Baker Stove** is designed to improve the efficiency of cooking in rural places.

Find out about the campaign to provide these stoves using **Baker stove**.

Research the design of the stove. **Evaluate** the design in terms of heat transfer.

Discuss: How does the stove improve on the efficiency of cooking using traditional methods? How might it improve the lives of the people who use it?

Outline: What factors will affect the use of the design?

How much energy can be used?
What is efficiency?
Will energy 'run out'?

EFFICIENT WORK

In our experiment to measure work done by some falling masses, the kinetic energy gained by the trolley was never quite the same as the potential energy lost by the weights. This difference is due to energy lost, usually in the form of heat – although sound is another way in which a machine can lose energy. The energy is considered to be 'wasted' because, once it has been emitted into the environment, there is no way we can get it back to use it again.

The amount of useful work done compared to the total energy input gives us the **efficiency** of the machine:

$$\text{efficiency } e = \frac{W_{out} \text{ (J)}}{E_{in} \text{ (J)}}$$

Since efficiency is a ratio of energies, it has no unit. W_{out} is always less than E_{in}, so efficiency always has a value $0 < e < 1$. Often, it is expressed as a percentage – just multiply by 100.

The more efficient a machine or process, the better.

ACTIVITY: Efficient or not?

■ ATL

- **Information literacy skills**: Access information to be informed and inform others
- **Critical-thinking skills**: Interpret data

The table shows some typical efficiency values of common machines or devices.

■ **Table 6.2** Typical efficiencies

Machine or process	Typical efficiency e (%) (approximate)
human walking	24%
bicycle	90%
internal combustion engine (petroleum)	25%
steam engine	7% (last used on railways) – 26% (experimental)
electromagnetic transformer	90–97%

Interpret the data. Decide on three categories for the machines: high, medium and low efficiency. Categorize the machines accordingly.

Research how one machine in each of the three categories works. **Share** your findings with other groups.

As a class, **summarize** the common features shared by the machines in each category. What do you think makes the least efficient ones so inefficient and the most efficient ones so efficient?

◆ Assessment opportunities

In this activity you have practised skills that can be assessed using Criterion A: Knowing and understanding.

ACTIVITY: The dream of perpetual motion

■ ATL

■ **Critical-thinking skills**: Evaluate evidence and arguments; Develop contrary or opposing arguments

When the physics of work and efficiency was first elaborated in the early 19th century CE, some thought that it would be possible to make a machine that would run forever once it had been started with an initial input of energy.

SEE-THINK-WONDER

Interpret the pictures in Figure 6.21 showing some of the designs made for such a 'perpetual motion' machine.

What do you see? What do the pictures make you think? What do they make you wonder?

Discuss: What was the energy input for the machine? What was the energy output? What would the machine do, in reality?

Imagine somebody was trying to sell you one of the machines as a source of energy for your home. Explain, using the physics of efficiency, why you would not buy one of the machines.

◆ Assessment opportunities

In this activity you have practised skills that can be assessed using Criterion A: Knowing and understanding

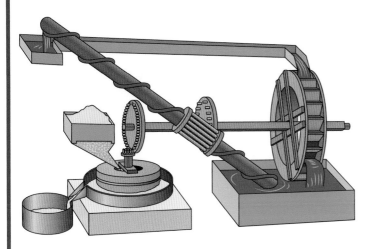
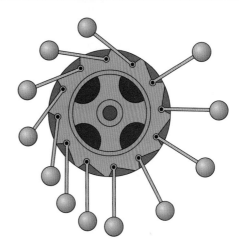

■ **Figure 6.21** Designs for perpetual motion

ACTIVITY: Investigating bounce energy

Use the investigation cycle (you can remind yourself of the features of the investigation cycle by turning to page 81) to design an investigation to find the energy loss in a bouncing ball. You will need to identify suitable variables to control and to measure, enabling you to estimate the energy lost as the ball bounces.

■ ATL

■ **Creative-thinking skills**: Use brainstorming and visual diagrams to generate new ideas and inquiries; Make guesses, ask 'what if' questions and generate testable hypotheses

◆ Assessment opportunities

This activity can be assessed using Criterion B: Inquiring and designing, and Criterion C: Processing and evaluating.

ACTIVITY: Machines that make a difference

A bicycle might seem a simple machine to many in the developed world. But a bicycle is a very efficient machine and it can have a huge impact on the quality of life of someone who, otherwise, has only foot power available.

Research some of the organizations that campaign and fund raise to send bicycles to developing parts of the world (here are a few examples):

- www.wheels4life.org/
- http://bicycles-for-humanity.org/
- www.worldbicyclerelief.org/our-work

Discuss: Why do these organizations think that bicycles can make such a big difference to lives of people in the developing world?

Research other examples of simple machines that can make a big difference to those in economically disadvantaged parts of the world. Some ideas might be:

- clockwork radios
- wind-powered water pumps
- solar-powered electricity systems
- gas-powered refrigerators.

Find out how the machines work and **explain** the energetic physics behind their operation.

Describe and **explain** how the machine you have chosen has an impact on human lives: what factors affect this impact? What challenges have to be overcome?

Be sure to refer to your research using **references** and **citations** where appropriate.

We can modify our energy-change diagrams to show the energy lost. One way to do this is to use a **Sankey diagram**. The arrows in the diagram are drawn such that their width is proportional to the amount of energy they represent at each stage of the process:

 Table 6.3 Energy changes in a wind turbine

Process	Energy (comparative units)
kinetic energy available from wind (example)	100
rotational kinetic energy in turbine blades	70
kinetic energy in turbine mechanism	45
electrical energy generated	20

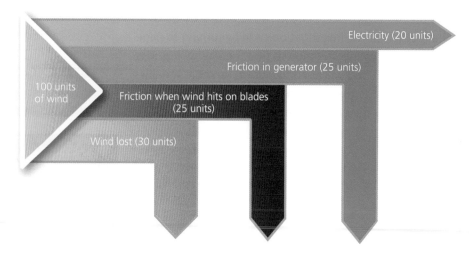

Electricity (20 units)

Friction in generator (25 units)

100 units of wind

Friction when wind hits on blades (25 units)

Wind lost (30 units)

■ **Figure 6.22** Sankey diagram for a wind turbine

Did you notice that the efficiency of a machine is defined in terms of the useful work out of the system? This, of course, involves a judgement as to what is, and what is not, useful. A microprocessor in a computer, in fact, wastes a lot of the input electrical energy in the form of heat, and this heat has to be dissipated using a heat sink and the fan you can usually hear whirring inside your computer.

Of course, we could use a microprocessor to heat our room when it is cold outside – but this would not be a very efficient use of the machine, since there are much easier ways to generate more heat than this.

■ **Figure 6.23** A heat sink is used to dissipate wasted energy as heat in a microprocessor

Sometimes it is useful to dissipate energy very quickly. A **crumple zone** is a specially designed area in a vehicle that crumples when the vehicle collides with another object. If the front of the vehicle were completely solid, all the energy of impact would be converted into deceleration and this would increase the likelihood of there being fatalities in the crash. By allowing the energy, instead, to deform the material on the front of the vehicle, the energy is dissipated slightly more slowly and so the violent decelerating effects of the impact are lessened.

ACTIVITY: CRASH!
Investigating crumple zones

■ **ATL**

■ **Creative-thinking skills**: Use brainstorming and visual diagrams to generate new ideas and inquiries; Make guesses, ask 'what if' questions and generate testable hypotheses

Research how crumple zones are used to improve vehicle safety. You may find these sites useful to begin with:

- www.euroncap.com/home.aspx
- www.howsafeisyourcar.com.au/

When a vehicle hits a stationary object, the kinetic energy of the vehicle is used to do work in the collision. If the front of the vehicle crumples by a distance, d, and this brings the vehicle to a complete stop, then

$$\text{work done } W = Fd = \frac{1}{2}mv^2$$

Use the investigation cycle to design an investigation into energy changed in a certain time for impacts of vehicles with different crumple zones.

◆ Assessment opportunities

This activity can be assessed using Criterion B: Inquiring and designing, and Criterion C: Processing and evaluating.

How is energy transformation controlled?

POWER TO THE PEOPLE

If you walk up a steep hill, you will feel tired at the top. The work done will be the energy changed, in this case the change in gravitational potential energy:

$$W = mg\,(h_2 - h_1)$$

What happens if you **run** up the hill? You will, most likely, feel more tired, since running means that you will do the same work in a shorter time. Changing our velocity means we are working at a different rate, but since the equation above does not include time, we need a different concept to understand the effect of doing work at a different rate. The physics concept of **power** is used to account for the difference in time:

$$\text{Power } P(\text{W}) = \frac{\text{Work done (J)}}{\text{time (s)}}$$

The power of a machine is measured in **Watts**, named after the experimenter, engineer and industrialist James Watt (1736–1819). Notice that **1 Watt = 1 Joule of work done in 1 second**.

For our example, the force against which we are doing work is the weight caused by the force of gravity, $F = mg$. If we generalize for any force using our work done equation,

$$P = \frac{Fd}{t}$$

$$v = \frac{d}{t}$$

So

$$P = Fv$$

■ **Figure 6.24** Specification plate on an electric kettle

Power is a very useful concept which applies equally well to mechanical machines as to electrical and other kinds of machines (more details in Chapter 10). For example, we can work out the work done (and so the energy required) to boil 1 litre of water using an electrical kettle. Take a kettle and look closely underneath it or on the side – you should see an information panel like the one shown in Figure 6.24.

The power is given usually in **kilowatts**. If you now fill the kettle with 1 litre of water, then time how long it takes to boil the water, you can use the power equation to work out the energy required, since

$$\text{work done} = \text{energy changed} = Pt$$

SOME SUMMATIVE PROBLEMS TO TRY

Use these problems to apply and extend your learning in this chapter. The problems are designed so that you can evaluate your learning at different levels of achievement in Criterion A: Knowledge and understanding.

THIS PROBLEM CAN BE USED TO EVALUATE YOUR LEARNING IN CRITERION A TO LEVEL 3–4

1 Figure 6.25 shows one early idea for a way to use horses to pull railway trains.

■ **Figure 6.25** Early design for a horse-powered railway locomotive

   a **Outline** the energy changes in this machine, including energy lost.
   b **Discuss** whether this locomotive was likely to be more, or less efficient than the horse alone.
   c One unit of power is the 'horsepower' (hp) where 1 hp = 746 W. A small steam locomotive can produce approximately 140 kW. **Calculate** how many horses would be required to produce the same power as the steam engine.

THIS PROBLEM CAN BE USED TO EVALUATE YOUR LEARNING IN CRITERION A TO LEVEL 5–6

2 The table below shows energy inputs and outputs for an internal combustion engine (gasoline) and for an electric motor.

Energy loss in an electric motor	
mechanical energy as sound	20%
heat energy by mechanical friction	5%
heat energy by electrical resistance	20%
Energy loss in an internal combustion engine (gasoline)	
mechanical energy as sound	25%
heat energy by mechanical friction	5%
heat energy in exhaust gases	30%
heat energy in cooling system	15%

   a Use the information in the table to **calculate** the percentage efficiency of each device.
   b **Summarize** the energetic performance of the two devices in the form of Sankey diagrams.
   c In a country, the average cost of electricity is approximately $0.10 per 3.5 MJ. The average cost of 1 litre of gasoline is $0.97. 1 litre of gasoline can produce 32 MJ of energy on combustion. **Calculate** the cost of 1 MJ of each energy source, and so **compare** the costs of running vehicles powered by electricity and vehicles powered by gasoline.
   d Many engineers believe that electric vehicles will be more environmentally friendly than internal combustion vehicles. **Discuss** this view with reference to your analysis of the information in the table.

3   Figure 6.26 shows a design for a fairground rollercoaster
ride.

■ **Figure 6.26**

At the beginning of the ride, the train is pulled to
the top of a steep incline whose peak is 10 m above the
starting point. Then the train is let loose and descends
the incline under gravitational force only. The train then
climbs a second incline that is height 5 m above the
starting point. The train has mass 500 kg.

a   Outline the energy changes taking place as the
rollercoaster completes the whole track.

b   Calculate the potential energy gained by the
rollercoaster in climbing to the top of the first incline
(point X). Assume $g = 10\,\text{ms}^{-2}$.

c   Calculate the theoretical velocity at point Y,
after descending from the first incline. **State** any
assumptions you make and assume $g = 10\,\text{ms}^{-2}$.

d   The velocity at point Y is in fact $v = 12\,\text{ms}^{-1}$. **Explain**
the difference between your calculated value and
the actual velocity.

The second incline is exactly half the height of the first,
and the slope of the second incline is constructed such
that the length of the track for the train to travel is also
exactly half that of the first.

e   Calculate the gravitational potential energy the train
must gain in climbing the second incline to point Z.

f   Assuming that energy is wasted in climbing from
point Y to point Z at the same rate as in descending
from point X to point Y, **show that** the velocity of
the train when it reaches point Z is $v = 9.43\,\text{ms}^{-1}$.

g   The rollercoaster engineer wants to add further
'bumps' to the ride, like the one between Y and Z.
Estimate the total amount of energy lost in this
bump, and so **suggest** how many further bumps
the train can climb before stopping. (Note that you
do not have to calculate an exact value here.)

h   Discuss the validity of the assumption made in
part **f** and **summarize** the effect on the design of
the rollercoaster of any different assumptions you
might make.

Reflection

In this chapter we have defined energy as the capacity to cause change. Where this change is useful, it is called **work done**. Machines are designed to control the way that energy is transformed as work, but all machines are wasteful – some of the input energy is always lost, usually as heat and sometimes as sound. We explored the interaction of heat with matter, considering the meaning of heat energy and what we actually measure when we measure temperature. We also looked at the three principal processes by which heat is transferred. Comparing the amount of useful work obtained from the machine to the energy lost in the process gives us the concept of **efficiency**. The efficiency of machines is always less than 100% as all energy transformations result in the generation of heat. Considering the rate at which work is done gives us the **power** of our machine.

We have reflected on the ways in which even simple machines can make a big difference to human lives, but access to those machines is not equally distributed across the world – some people live daily surrounded by machines, while others still have to rely heavily on the resource of their own body.

Use this table to reflect on your own learning in this chapter.		
Questions we asked	Answers we found	Any further questions now?
Factual: What are examples of energy? What is temperature? How can we measure the work done? How does heat energy transfer? How can we measure the efficiency of machines?		
Conceptual: What is energy? How is energy harnessed to do work? How does energy interact with matter? How is energy transformation controlled? How much of the energy available can be used to do useful work?		
Debatable: Do the machines we make always change things for the better? Will energy ever 'run out'? Can a machine be made that will never run out of energy?		

Approaches to learning you used in this chapter	Description – what new skills did you learn?	How well did you master the skills?			
		Novice	Learner	Practitioner	Expert
Information literacy skills					
Communication skills					
Critical-thinking skills					
Creative thinking skills					
Transfer skills					
Learner profile attribute(s)	Reflect on the importance of being caring for our learning in this chapter.				
Caring					

7 How can we communicate?

○ New global *relationships* have become possible as humanity has learned to *communicate* through *energy* transferred as wave *motion*.

CONSIDER AND ANSWER THESE QUESTIONS:

Factual: How do we measure a wave? What affects the speed of a wave? What is sound? What is light? What other kinds of wave are there? How can we manipulate and control light waves?

Conceptual: How does wave motion differ from other kinds of motion? How do we experience different kinds of wave? Where is the kinetic energy in wave motion? How can a wave carry information? How does our experience of waves affect the quality of our communication? How has improved communication affected our world?

Debatable: To what extent have improved communications made the world a 'global village'? What advantages and disadvantages might global telecommunications bring?

Now **share and compare** your thoughts and ideas with your partner, or with the whole class.

■ **Figure 7.1** What happens between the headphone speaker and your brain?

○ IN THIS CHAPTER, WE WILL ...

■ **Find out** how we communicate using kinds of wave energy, how waves move, and how they are affected by what they move through.

■ **Explore** how our personal experience of different kinds of wave energy is related to the form of the waves themselves, and how we can use wave energy to communicate better.

■ **Take action** to communicate and share our learning with other MYP schools, and use physics as a way to express ourselves artistically.

- Critical-thinking skills
- Creative-thinking skills
- Communication skills
- Transfer skills

● We will reflect on this learner profile attribute …

- Communicators – we will explore how the physics of waves enables us to communicate better and espress ouselves in different ways

◆ Assessment opportunities in this chapter

- **Criterion A**: Knowing and understanding
- **Criterion B**: Inquiring and designing
- **Criterion C**: Processing and evaluating
- **Criterion D**: Reflecting on the impacts of science

What do the objects in Figure 7.2 have in common?

You might have answered that they are all designed for communication. You might also have answered that they all produce waves of one kind or another. Human beings have evolved to experience the world, and communicate with each other, through the use of two particular kinds of energy that travels as waves – sound and light. As a consequence, we have developed ways to refine that communication and even turn it into forms of creative expression, such as music. We have also extended the ways we communicate to include new kinds of wave – especially microwaves and radio waves.

ACTIVITY: Observing sound makers

- ATL

- **Critical-thinking skills**: Practise observing carefully in order to recognise problems

Watch sound sources in action at the following websites:

String instrument oscillations – guitar, violin

> https://www.youtube.com/watch?v=6JeyiM0YNo4

Endoscope singing vocal cords

> https://www.youtube.com/watch?v=Tsh-0VvAGQA

> https://www.youtube.com/watch?v=-XGds2GAvGQ

While watching, **think** about these guiding questions, **pair** and **share** your answers with the class.

- **What is physically making the sound?**
- **How is this object making the sound? Describe the motion it makes.**

KEY WORDS

longitudinal	prism
medium	ray
media	signal
oscillation	transverse

- **Figure 7.2** Guitar, drum, vocal cords, laser and optical fibre, radio transmitter

How does wave motion differ from other kinds of motion? How do we measure a wave?

WAVE HELLO!

■ **Figure 7.3** Ripples made by a water droplet

Have you ever watched ripples spreading out from a water droplet? The ripples move outwards from the centre point, but the water itself does not move in that direction – if it did, we would end up with all the water at the edges of the pool and none left in the middle. Nevertheless, the ripples **are** moving – you can see them! – and are carrying kinetic energy through the water from the starting point, or **origin**, of the motion where the water droplet hit the surface of the pool.

The water, in this case, is the **medium** for this special kind of motion, and the phenomenon by which energy can move through a medium without causing any net displacement of the medium itself is called **wave motion**.

In the 'Activity: Observing wave motion', in each case the spring is moving 'to and fro', and this motion is **propagated** through the rest of the spring – so the kinetic energy of the original **disturbance** is transmitted as a wave through the medium.

■ In the first case, the disturbance is moving at right angles to the direction in which the energy moved through the medium. This is called a **transverse oscillation**.
■ In the second case, the disturbance is moving in the same direction as the energy moving through the medium. This is called a **longitudinal oscillation**.

As the wave passes by, the medium moves back and forth around its original position. This motion is called **oscillation**.

A **pendulum** is any mass swinging about a centre point on a string of some kind. For example, a child's swing is a pendulum. Pendulums were of great interest to early physicists, including Galileo Galilei (1564–1642). There is a story that Galileo started thinking about how useful the regular motion of a pendulum might be, while bored in church one day. The story goes that he was watching the great incense burner of the cathedral of Pisa swinging to and fro over the congregation, and he decided to pass the time by measuring how long it took the burner to swing using his own pulse.

Pendulums were later used as accurate timing devices, including in the first mechanical clocks. Until then, time had been measured by less reliable means, such as the burning down of a candle or the running of sand through a hole.

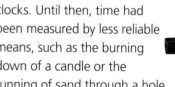

■ **Figure 7.4** Pendulum clock

ACTIVITY: Observing wave motion

In this activity you will use a slinky spring to observe wave motion of different kinds.

In pairs, take one end each of a slinky spring and lay it straight along a smooth surface, such as a corridor floor.

Decide who will be the 'source' of the motion, and who will be the 'observer'.

We are beginning to gather information to aid our inquiry by making **preliminary observations**. From these observations we aim to **characterize** the nature of wave motion. This is often the first stage of any scientific inquiry.

1 The 'source' should flick the spring rapidly up and down, perpendicular to the spring's axis. **Observe** the disturbance as it propagates through the spring.

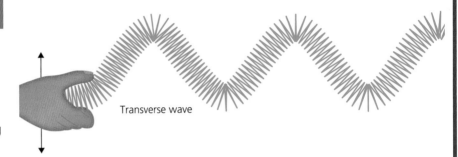

Transverse wave

■ **Figure 7.5** A slinky spring being oscillated perpendicular to its axis

2 When the spring has become still again, the source should now give the spring a quick, hard push along its axis. **Observe** the disturbance as it propagates through the spring.

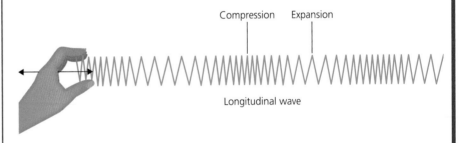

Compression   Expansion

Longitudinal wave

■ **Figure 7.6** A slinky spring being oscillated parallel to its axis

ACTIVITY: Up and down, to and fro, side to side

Watch some examples of extreme wave motion. In YouTube, search the following Tacoma Narrows Bridge; Millenium Bridge London opening; 1989 San Francisco earthquake and East Japan eathquake 2011.

Discuss and decide whether the motion you are observing is transverse or longitudinal. **Describe** the evidence are you looking for in each case, and **organize** your observations in a table with these headings:

Observation	Evidence for ...	Explanation

ACTIVITY: There and back again

Work in pairs.

Aim: To investigate the factors affecting the time for one swing of a pendulum.

Background: If we want to use a pendulum as a timer, we need to know the factors that affect the time of its swing. The time of a swing is called the **time period**, *T*.

Possible factors might be:
- **length of the string**
- **mass of the pendulum**
- **initial height of the pendulum**.

Hypothesis

Choose a variable from the list and **write a hypothesis** about how you think it will affect the time period, *T*. **Explain** your hypothesis with scientific reasoning about the forces and the motion of the pendulum.

Method

- **Independent** variable to change: state your own
- **Dependent** variable to measure: time period
- **Variables to control**: state your own

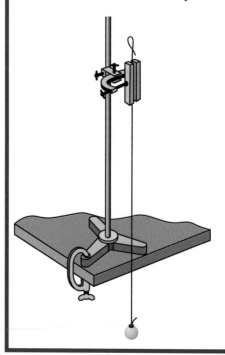

Suspend the pendulum from a lab stand. Position the pendulum so that the origin point – when the pendulum is hanging straight down and stationary – is in line with the lab stand when viewed from the front:

The time period for one oscillation of the pendulum is the time taken for the pendulum to travel back and forth and then return to its starting point again.

Discuss: Will the pendulum motion remain the same all the time? What will happen to the pendulum's motion over many swings? How will you control the variables you are not changing? How will you make sure that you measure the time period for one swing accurately?

Outline a method that will allow you to measure the time period for the pendulum' swing accurately.

Results

Design a suitable results table clearly showing units of measurement.

Analysis

The time period of the pendulum is not related linearly to any of the variables! However, there is a relationship between the square of the time period, T^2, and one of the variables. **Present** your data on a suitable graph to test the relationship between T^2 and your variable.

Outline the significance of the gradient of your graph.

Conclusion

Write a conclusion. Was your hypothesis verified, or falsified? **Explain** your findings by researching an equation for time period of a pendulum, and with reference to the physics of forces and motion.

Evaluation

Evaluate your results with reference to the equation you found from research, and the graph that you obtained. Was there an intercept on your graph? **Comment** on what this might mean.

Calculate the overall percentage error in your results.

Comment on the sources of error, and **compare** their likely effect on the final result.

■ **Figure 7.7** Set-up for pendulum experiment

In order to compare waves, we can measure their dimensions:

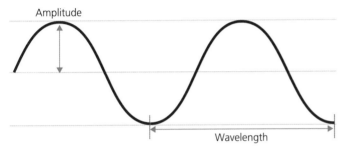

■ **Figure 7.8** Dimensions of a wave

Note that the **amplitude** of the wave is the maximum displacement from the starting point or 'equilibrium position' – **not** the distance between the top and bottom of the wave. Similarly, the **wavelength** is the distance from one point on the wave to **the same point on the next wave** – so this could be from peak to peak, or trough to trough, for example.

How do we measure the rate at which the oscillation is occurring? We can define the **frequency** as the number of oscillations that occur every second, or

$$\text{frequency } f \text{ (Hertz)} = \frac{1}{T(s)}$$

The unit of frequency is, thus, the number of oscillations per second, usually called the 'Hertz' after the German radio pioneer Heinrich Hertz (1857–1894.)

If we think of a wave as stretched out in space (like the slinky spring waves or waves on water) then the wave will travel one wavelength, λ, in the time it takes to complete an oscillation – which is the time period, T. From this, we can use the kinematics equation for velocity (see Chapter 4) to obtain the velocity of the wave through the medium:

$$v = \frac{d}{t} = \frac{\lambda}{T}$$

But from our equation for frequency we can substitute $f = \frac{1}{T}$ and

$$v \, (\text{ms}^{-1}) = f(\text{Hz})\lambda(\text{m})$$

If λ is measured in metres and f in Hertz, then the velocity of the wave will be in ms⁻¹.

The speed of the wave will depend on the way the medium allows the wave to pass or **propagate**.

ACTIVITY: Analysing wave speeds

ATL

- **Critical-thinking skills**: Draw reasonable conclusions and generalizations

Work individually or in pairs.

Table 7.1 shows the speed of sound through various different media.

- **Table 7.1**

Medium	Speed of sound (m s⁻¹)
vacuum	does not travel
air (20°C)	343
air (0°C)	331
water (25°C)	1493
glass	5640
ground (granite rock)	5950

1 **Describe** the effect on the speed of sound of the different materials. **Suggest** what property of the materials is important in determining the speed of sound.
2 **Explain** why this property should affect the speed of sound.

A shaky problem

Some earthquake waves can be considered a kind of sound wave that is moving through the ground. A particular earthquake begins with its epicentre 100 km from the Mexican coast, in the Pacific Ocean. The earthquake lasts 5 seconds.

Some of the earthquake energy is propagated as waves through the seabed, while other energy is propagated through the sea itself. The first earthquake waves are detected at the Mexican coast 17 seconds after the earthquake begins.

3 **State** what medium the first waves must have travelled through.

4 **Calculate** when the second 'aftershock' waves will arrive at the coast.
5 **Calculate** how long the second 'aftershock' waves will last.
6 **State** which will probably have the greatest amplitude: the first waves or the aftershock waves? **Explain** your answer.
7 **State** an assumption we have made about the paths taken by the first and aftershock waves, and **explain** the significance of this assumption for the calculations.

◆ Assessment opportunities

In this activity you have practised skills that are assessed using Criterion A: Knowing and understanding.

- **Figure 7.9** Supersonic aircraft travel faster than the speed of sound. The speed of sound at a given altitude is called the Mach number

HEARING THE MUSIC AND SEEING THE LIGHT

Sound is a longitudinal wave. Sound generators cause variations in air pressure which propagate through the air and which are then detected as minute vibrations in our ear drum (Figure 7.10).

If we draw a line through all the regions of maximum or minimum pressure, we would obtain the crests or troughs of the waves – called **wavefronts**. The wavelength for a sound wave might then be measured as the distance between successive maximum or minimum wavefronts.

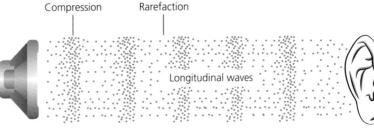

Compression   Rarefaction

Longitudinal waves

■ **Figure 7.10** Propagation of longitudinal pressure waves of sound

---

ACTIVITY: Experiencing sound waves

■ ATL

- **Critical-thinking skills:** Practise observing carefully in order to recognize problems

The form of a wave will affect the way we experience it. For sound waves, our ears directly distinguish different wavelengths, frequencies and amplitudes.

Work as a class.

Aim: To measure the audible range within the class.

Background

The range of frequencies we can hear depends on many factors, for example our age.

Equipment
- **Signal generator with adjustable amplitude, frequency**
- **Amplifier (unless this is built into the signal generator)**
- **Loudspeaker**
- **Datalogger or digital oscilloscope connected to the signal generator**

Your teacher will connect the apparatus. The class should position themselves in a large circle so that everybody is about 2 m distance from the loudspeaker.

Experiment 1: How do we experience the properties of sound waves?

Your teacher will produce a fixed frequency of 256 Hz but adjust the amplitude. Observe the effect on your experience of the wave, and on the trace of the wave on the oscilloscope.

Now your teacher will change the frequency. Observe the effect on your experience and on the oscilloscope.

Experiment 2: Audible cut-off

Your teacher will now disconnect the oscilloscope. The amplitude of the signal will be kept constant at all times, but your teacher will slowly change the frequency, starting at very low frequency.

Raise your hand when you can hear the sound, and keep it raised until you can no longer hear the sound.

What happens to the perceived volume of the sound?

Count and record the number of people with their hands raised every increment of 100 Hz.

Note the frequencies of wave at which you (a) just begin to hear it and (b) cease to hear it.

Analysis

Choose a suitable way to present your data, such that the audible range for the class can be measured.

◆ Assessment opportunities

In this activity you have practised skills that are assessed using Criterion C: Processing and evaluating.

Our ears are not equally sensitive to all frequencies. In the graph in Figure 7.11, the intensity of sound is measured in decibels (dB) against the frequency of the sound (Hz). Note that the decibel scale is not linear, but 'logarithmic.' This means that an increase of 10 dB corresponds to 10× the power, or approximately 3× the amplitude of sound.

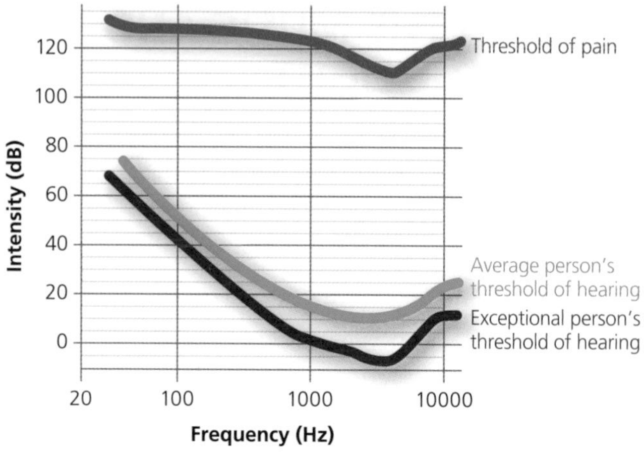

As a consequence of varying sensitivity, we think that we hear higher pitched sounds more quietly. Of course, other animals also have ears and can experience sound. However they do not experience the **same** sounds that we do.

So, what about light? Light is a whole different story... whereas other waves transfer kinetic energy through the motion of a medium, light is a kind of energy that can transfer 'itself' through space – in other words, light does not need a medium at all and can travel through a vacuum!

This was not something that was easy to accept: for a very long time, physicists were trying to detect the 'medium' through which light moved – which they called the ether, after the perfect 'fifth' element in the model of nature of the classical Greek philosopher Aristotle.

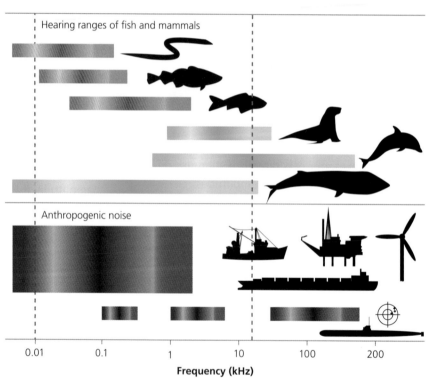

■ **Figure 7.11**
Indicative human hearing response with frequency

■ **Figure 7.12** Audible range for different animals compared to the noise produced by human activity (anthropogenic)

The American physicist A.A. Michelson spent most of his working life trying to show that the speed of light varied when it was travelling with, or against, the flow of the ether past the orbiting Earth. In 1887 Michelson carried out an experiment with E.W. Morley that was so accurate it ought to have been able to measure this difference. But it failed to do so. Michelson was never quite able to accept that the ether **did not exist**, but in his attempts to detect the ether he had measured the speed of light with ever greater accuracy, and he was compensated for his efforts by the Nobel Prize for Physics in 1907.

The work of the British physicist James Clerk Maxwell (1831–1879) in electricity and magnetism enabled physicists to understand light as an electric field and a magnetic field oscillating in space. The field strength is changing at right angles to the direction of propagation of the wave, so we regard light as a **transverse** wave (Figure 7.13).

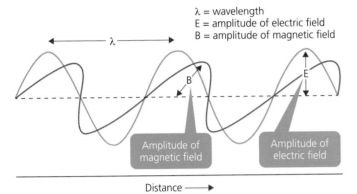

λ = wavelength
E = amplitude of electric field
B = amplitude of magnetic field

Amplitude of magnetic field

Amplitude of electric field

Distance ⟶

Figure 7.13 Light is a transverse oscillation of electric and magnetic fields at right angles to each other

Of course we experience light with our eyes. The **amplitude** of a light wave is related to the brightness or **intensity** of the light. The **frequency** of light is experienced as colour for visible light. However, it was soon clear to physicists that many other kinds of energy were transferred in the same way as light, through oscillating electromagnetic (EM) fields. In Chapter 5 we encountered the phenomenon of heat radiation. Heat can transfer through a vacuum in the form of **infra-red** radiation, which we do not 'see' as such

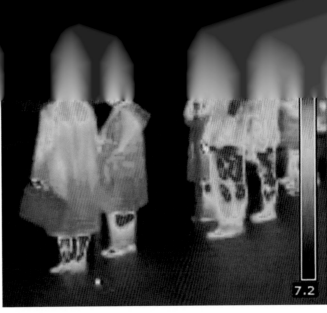

7.2

Figure 7.14 Infra-red topographic photograph of a human body. The colours are added artificially to represent the different intensities of infra-red radiation being released.

but which we experience as a heating effect in matter, such as our skin.

Different creatures are capable of experiencing still other frequencies of EM radiation. Pollinating insects can often 'see' high frequency ultra-violet radiation, since this enables them to identify flowers that reflect these radiations very strongly. Similarly, rattlesnakes can 'sense' the infra-red radiation produced by their prey.

Physicists have classified the kinds of EM radiation in terms of their frequency and their effect on matter. This classification is then represented as the electromagnetic or EM **spectrum** (Figure 7.15).

We will look further at the EM spectrum in the context of space and observations in Chapter 12.

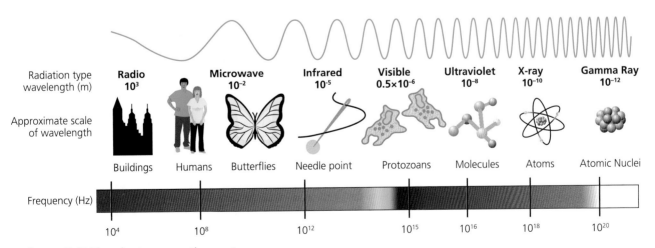

Figure 7.15 The electromagnetic spectrum

How can a wave carry information? How can we manipulate and control light waves?

HOW DO WAVES HELP US TO COMMUNICATE?

Fortunately for the evolution of human society, nature has equipped us very well for short-range communication; we can manipulate sound with great subtlety to produce language, and we have very sensitive light detectors to enable us to see to whom we are speaking! If we cannot hear, we can use our eyes to observe signs for communication instead (Figure 7.16).

Communication over longer distances is more problematic. A gust of wind can snatch our speech away, while a foggy day could obscure any attempts at visual communication.

One successful communication system was developed by the American portrait painter and inventor Samuel Morse (1791–1872) in 1837. Morse's system had the advantage of being robust and his Morse code was more reliable as it had three 'states' for transmission of information: a long signal or 'dash', a short signal or 'dot', and no signal.

▼ Link: Music

Music to our ears!

■ ATL

■ **Transfer skills:** Make connections between subject groups and disciplines

Of course, the physics of sound waves is very closely connected to music. For example, if you play a musical instrument or can read music, you may be familiar with the musical concept of the 'octave' in the Western tradition. An octave is the interval between the same note and its repetition in the Western musical scale. For example, the frequencies for the notes beginning a Western scale at 'middle C' are shown in Table 7.2.

■ Table 7.2

	Scale step (semitones)	Approximate frequency (Hz)
C	2	262
D	2	294
E	1	330
F	2	349
G	2	392
A	2	440
B	1	494
C		524

An octave in music represents the doubling of frequency of whatever note you started out at. The steps in the scale are determined also by ratios of frequencies between notes.

What

name

is

your

My

I

am

happy

■ **Figure 7.16** Sign language

Morse code is not a wave as such – the electric current is at a fixed voltage and is simply on or off for certain amounts of time. Later, waves became useful as a way to transmit information with no physical connection between transmitter and receiver at all. Heinrich Hertz showed experimentally in 1887 that energy could be transmitted from an oscillating electrical signal in a coil to a circular antenna nearby. His intention was actually to test James Clerk Maxwell's predictions about electromagnetic waves (mentioned earlier in this topic), so when asked whether he thought there were any useful applications for his discovery Hertz replied, 'Nothing, I guess.' He had, in fact, discovered radio waves!

Others were quicker to see the potential of Hertz's discovery and experiments continued throughout the 1890s. The first practical system for transmitting information using radio waves was probably developed by the Russian researcher and engineer Alexander Popov (1859–1906) in 1895, although Guglielmo Marconi (1874–1937) achieved a similar result towards the end of 1895 in Bologna, Italy. In any case, Marconi was the most energetic at patenting and commercializing his discoveries, and the development of radio was driven by his Marconi Company for some time afterwards.

At first, the messages sent by radio were in Morse code. During the early 20th century CE, however, scientists and engineers began to realize the potential for transmitting the human voice by radio. This required a more sophisticated system for coding the information, called **amplitude modulation** (AM) transmission. The voice is not converted directly into a radio wave, since the radio wave would then have to have the same frequency range as the voice. Instead, a **carrier wave** is used with a fixed frequency. The amplitude of **this** wave is then varied or 'modulated' by the information you want to transmit (Figure 7.17). Some radio stations still transmit using AM today.

Radio waves can travel (to an extent) through buildings, hills and other obstacles, and – at lower frequencies – they are reflected by the Earth's **ionosphere**, which behaves like a mirror in the sky, so that long-range communications are possible. AM transmission, however, is still somewhat limited in terms of the quantity of information that the carrier can convey. **Frequency modulation** (FM), instead, uses a varying frequency of carrier wave to transmit information and this produces enough **bandwidth** for high-quality stereo sound (Figure 7.18).

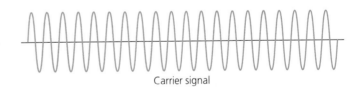

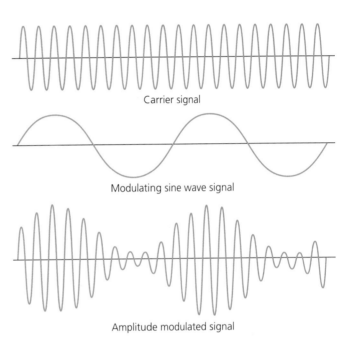

■ **Figure 7.17** Amplitude modulation of sine waves

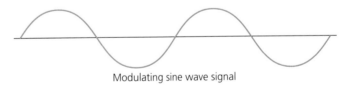

■ **Figure 7.18** Frequency modulation of sine waves

The radio region of the EM spectrum is now divided up and certain frequency 'bands' or channels are licensed for use by governments. Similarly, mobile telephones (cell phones) and Wi-Fi devices use the much higher frequency **microwave** region to communicate with the stations within a cell network or with each other.

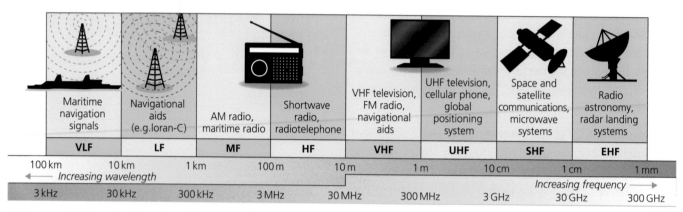

Maritime navigation signals	Navigational aids (e.g.loran-C)	AM radio, maritime radio	Shortwave radio, radiotelephone	VHF television, FM radio, navigational aids	UHF television, cellular phone, global positioning system	Space and satellite communications, microwave systems	Radio astronomy, radar landing systems
VLF	**LF**	**MF**	**HF**	**VHF**	**UHF**	**SHF**	**EHF**

| 100 km | 10 km | 1 km | 100 m | 10 m | 1 m | 10 cm | 1 cm | 1 mm |

← *Increasing wavelength*

| 3 kHz | 30 kHz | 300 kHz | 3 MHz | 30 MHz | 300 MHz | 3 GHz | 30 GHz | 300 GHz |

Increasing frequency →

■ **Figure 7.19** Detail of bands within the radio region of the EM spectrum

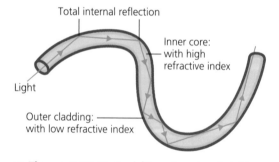

Total internal reflection

Inner core: with high refractive index

Light

Outer cladding: with low refractive index

■ **Figure 7.20** Optical fibre showing total internal reflection

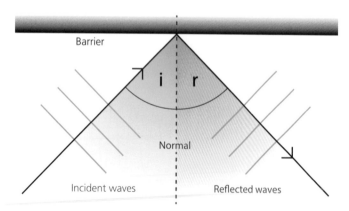

Barrier

i r

Normal

Incident waves

Reflected waves

■ **Figure 7.21** Wavefront diagram for reflection

Ironically, the most recent developments in telecommunications technology have gone **back** to the idea of hooking up devices with a cable and using a binary, 'on–off' communications protocol. The difference is that these systems do not use electric signals or even radio waves, but use light to carry the information. An **optical fibre** carries beams of laser light that are amplitude modulated to represent 'bits' of information – but how is light used to carry this information?

The two most important phenomena for this technology are **reflection** and **refraction**. We can visualize these effects in terms of the 'wavefronts' (lines drawn along the peaks or troughs of waves) discussed earlier (page 116).

▼ Link: History

■ ATL

■ **Transfer skills:** Make connections between subject groups and disciplines

Hertz and Popov invented radio communication, but it was Marconi who made it work and brought it to the world. Who is most important in history: the inventor, or the communicator? The originator, or the popularizer?

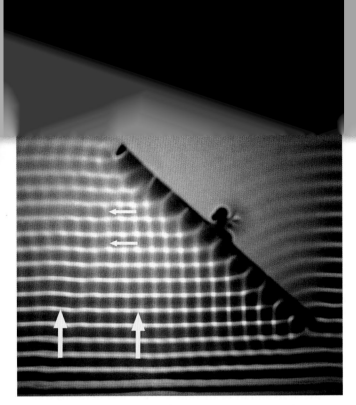

■ **Figure 7.22** Reflection of water waves as seen in a ripple tank

As each wavefront meets the reflecting barrier, its direction is changed and the energy of the wave is directed back away from the barrier. We can represent the direction of energy flow by drawing lines along the middle of the wavefronts – this simplifies the picture into a **ray diagram**.

Refraction occurs where the light passes into the medium (because it is transparent) and continues – but the direction of the ray is changed because the speed of the wave in the medium is different.

The quantity

$$n = \frac{c\ (\mathrm{m\,s^{-1}})}{c_{medium}\ (\mathrm{m\,s^{-1}})}$$

is called the **refractive index** of the medium, where c is the speed of light in a vacuum, which is very close to $3 \times 10^8\ \mathrm{m\,s^{-1}}$ – and is the same for all kinds of EM radiation. Since n is a ratio of two velocities, it has no unit. The value c is always the same, and is thought to be the maximum possible velocity that can be attained in the Universe. Note, however, that the speed of light **does** change and, indeed, is much slower in many media (Table 7.3).

■ **Table 7.3**

Speed of light ($\times 10^8\,\mathrm{m\,s^{-1}}$)	Medium
2.998	vacuum
2.997	air (20°C)
2.997	air (0°C)
2.254	water (25°C)
2.039	Perspex™
1.972	crown glass

What is more, although all EM radiation of all wavelengths travels at c in a vacuum, the speed of different wavelengths of EM radiation is different in different media … in other words, **refractive index depends on wavelength**.

ACTIVITY: Manipulating light

■ ATL

■ **Critical-thinking skills**: Interpret data; Evaluate evidence and arguments; Draw reasonable conclusions and generalizations

Aim: To investigate the properties of reflection and refraction of light.

Background

We already know that the speed of waves is affected by the medium through which they are travelling. When the medium changes, we can observe the effects caused.

Equipment
- Semi-circular transparent prism (plastic or crown glass)
- A ray box with single slit
- Millimetric graph paper, ruler, sharp pencil
- Angle measurer

Variables
- **Independent** (to change): angle of incidence of the light ray
- **Dependent** (to measure): angle of reflection and angle of refraction of the light ray
- **Controlled**: intensity of light ray, frequency(ies) in light ray, media for incident and transmitted rays

Procedure
Set up the apparatus as shown in Figure 7.23. ➤

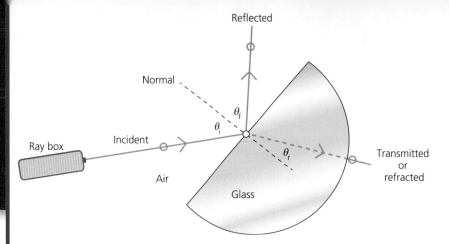

■ Figure 7.23 Reflection/refraction experiment

Hypothesis

How will the incident ray be affected by the prism? What will happen as the prism is moved so that the rays enter the prism at a different angle? **Explain** your answer.

The **ray box** is an intense source of white light that can be shielded or 'blinded' with card inserts. Use a card insert that has a single slit so that the light emerges from the ray box as a single ray.

Mark on the graph paper as follows:
- Draw along the straight edge of the prism so that you can record the position. Mark this line with a number '1'.
- Carefully mark dots or small dashes along the centre line of the incident ray. Mark this row of dots with a '1' also.
- Look for a ray that reflects back where the incident ray hits the first outside surface of the prism. Mark along the centre line of this ray with dots and number it '1A'.
- Look for where the ray emerges from the other side of the prism. Mark along the centre of this ray with dots also and number '1B'.
- Now, turn the prism a few degrees around its centre point, without moving the ray box or incident ray. Repeat the markings as above and number prism position 2, reflected ray 2A, transmitted ray 2B, etc. Continue until you have a range of results.

Analysis

The diagram shows how to measure the **angle of incidence** θ_i, **angle of reflection** θ_f, and the **angle of refraction** θ_r.

Record these angles in a suitable table.

Present your data on suitable graphs to show the effects of reflection and of refraction. **Outline** any patterns you see in the data.

Conclusion

State the relationship between **angle of incidence** θ_i and **angle of reflection** θ_f.

For the **refracted** rays, it can be shown that

$$\frac{\sin \theta_i}{\sin \theta_r} = \frac{c_1}{c_2}$$

where c_1 and c_2 are the velocities of light in the first medium and the second medium respectively.

Research to find the speed of light in air, c_1.

Use your data or a suitable **linearized** graph to **calculate** the speed of light in the prism c_2.

Evaluation

Research this value for the prism material you have been given and **compare** to your experimental value. Give the overall percentage error in your result, and **discuss** the contributing sources of error and their relative significance.

Suggest ways to improve the experiment such that these sources of error might be reduced.

ACTIVITY: Seeing colours

Aim: To measure the bandwidth of coloured filters.

Background

A coloured filter is used to absorb selected wavelengths from incident light and transmit the remainder. This reduces the range of wavelengths in the transmitted light, and this range of wavelengths is called the **bandwidth**.

A prism or a diffraction grating can both be used to separate out the colours in incident light of many wavelengths, such as white light. They separate out the different wavelengths in the incident light spatially to make a **spectrum**.

Variables

- **Independent**: colour of filters – red, yellow, green, blue, violet
- **Dependent**: bandwidth of transmitted light
- **Controlled**: intensity, wavelength composition of incident light

Procedure

Set up the apparatus as shown in Figure 7.24. The diffraction grating or the prism can be placed at point X.

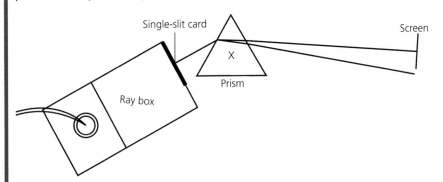

Single-slit card

 Screen

X

Prism

Ray box

■ **Figure 7.24** Experiment setup for measurement of bandwidth of filters

You should carefully adjust the angle of incidence of the initial light ray to find the transmitted beam on the screen. **Do not** change the distance of the prism/diffraction grating to the screen.

Use a pencil to mark off the beginning and end of the transmitted spectrum for each filter. Write down any colours visible in the transmitted spectrum.

Analysis

Since the angle of diffraction or refraction depends on the wavelength of the incident light, the width of the spectra you see on the screen is proportional to the bandwidth of the colours transmitted. **Outline** the bandwidths observed for each of the colours and **comment** on the different bandwidths you have measured.

◆ Assessment opportunities

In this activity you have practised skills that are assessed using Criterion C: Processing and evaluating.

ACTIVITY: A smaller world?

It is commonplace for people to talk about a 'global village' where everyone can communicate with everyone else. But how true to reality is this view of our world?

Research one communications technology to find out how it works. Choose one that enables you to apply the knowledge and understanding of waves gathered during this topic. Write a magazine article about your chosen communications technology. Your article should be written for students in another school who are the same age as you, but who do not study physics.

- **What problem was your chosen communications technology expected to solve? Explain how this was achieved.**
- **What other factors affected this communications technology? What other effects did the communications technology have? For example, social effects, economic limitations or developments, challenges to our ideas about right and wrong, public or private …**
- **Use correct scientific terminology throughout your report, and cite and reference your research sources.**

Why not share all of the articles from your class online, perhaps on a blog or similar with the heading 'How small does communications technology make our global village?' If you know how, add a comments section and invite students in other MYP schools to give feedback.

ACTIVITY: Waves of applause

Use what you have learned in this chapter to put together a class show for an assembly, town meeting or as a science fair event. The show should demonstrate to others the physics that lies behind a particular kind of art: whether it be performing art or visual art. Work together in small groups to elaborate different events.

Some ideas:

- **perform with musical instruments and oscilloscopes**
- **the physics of my music – explore musical scales and musical forms from different cultures and find out about the physics of waves behind them**
- **write and perform a 'silent song', audible only to dogs! Then perform it so that humans can hear it!**
- **use filters and spectra to create a light show**
- **use soap bubbles to create coloured light-sculptures.**

Whatever you do, express yourself!

SOME SUMMATIVE PROBLEMS TO TRY

Use these problems to apply and extend your learning in this chapter. The problems are designed so that you can evaluate your learning at different levels of achievement in Criterion A: Knowledge and understanding.

THIS PROBLEM CAN BE USED TO EVALUATE YOUR LEARNING IN CRITERION A TO LEVEL 3–4

1 A surfer is watching the waves arrive on her favourite beach in a bay. She wants to know how quickly the waves will carry her into the beach on her surfboard.

   a **State** two variables the surfer will need to know in order to work out the velocity of the waves.

   The surfer compares the distance between wave crests to the distance between two trees that the waves are passing on the side of the bay. The distance between these trees is about 10 metres. She also notices that a boat in the bay is bobbing up and down as the waves pass. For the boat to bob to its highest point each time takes about 6 seconds.

   b **Interpret** the surfer's observations to **calculate** the velocity of the water waves.

   c The waves seem to get higher as they come closer to the shore, where the water is shallower. **Outline** why this might be happening.

THIS PROBLEM CAN BE USED TO EVALUATE YOUR LEARNING IN CRITERION A TO LEVEL 3–4

2 An antique clock uses a pendulum to keep time. The pendulum swings to make the clock 'tick' once every second.

   a **Calculate** how long the pendulum should be (assume gravitational acceleration $g = 10\,\mathrm{m\,s^{-1}}$).

   The owners of the clock move to a hotter country, nearer to the equator. They think that the clock is ticking more slowly now, for some reason.

   b **Suggest** one or more reasons why the clock might be running slower. **Outline** why this should be so.

One of the owners thinks they can compensate for the slower tick by making the pendulum 'bob' a little heavier. They stick some putty onto the bob to test this hypothesis.

   c **State** whether this will solve the problem of the slower tick and **explain** your answer with scientific reasoning.

THIS PROBLEM CAN BE USED TO EVALUATE YOUR LEARNING IN CRITERION A TO LEVEL 5–6

3 A fibre optic works because the laser light that is transmitted through the centre of the fibre becomes 'trapped' inside the fibre through reflection from its walls. The 'trapping' of the light in this way is called total internal reflection. Light will be reflected from a surface when its angle of incidence is greater than the critical angle given by

$$\theta_c = \sin^{-1}\frac{n_2}{n_1}$$

   Where n_2 is the refractive index of whatever is outside the fibre, and n_1 is the refractive index of the fibre medium itself.

   The table shows the speed of light in some different materials.

Speed of light ($\times 10^8\,\mathrm{m\,s^{-1}}$)	Medium	Refractive index n
2.998	vacuum	1.000
2.997	air (20°C)	
2.997	air (0°C)	
2.306	silicon dioxide glass	
2.254	water (25°C)	
2.039	Perspex™	
1.972	crown glass	

   a Use the information to complete the table by **calculating** the refractive index, n, of each of the materials.

   b Thus **calculate** the critical angle for total internal reflection to occur for an optical fibre made from crown glass in air at room temperature.

---

c **Describe** the effect on the critical angle of placing the optical fibre in water.

d **Outline** why the effect on critical angle you described might be a problem for communications using the fibre.

e An engineer suggests that a solution to this problem might be to 'wrap' the core of the fibre (the part carrying the signal) with a material of lower refractive index. **Show** the effect of doing this with a suitable material from the table and **explain** whether this suggestion will work.

THIS PROBLEM CAN BE USED TO EVALUATE YOUR LEARNING IN CRITERION A TO LEVEL 7–8

4 The data in Figure 7.25 is for the **strength of the radio signal** from a cell phone over different **distances** from the phone.

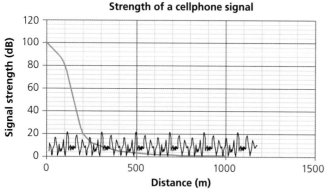

Strength of a cellphone signal

■ **Figure 7.25** Signal from a cell phone showing background noise

On the graph, two sets of data are shown:
- the cellphone signal strength
- the random 'noise' produced by other sources.

a **Annotate** the graph to show the different data.

b **Describe** how the signal strength varies with distance. **State** whether the relationship is proportional or not.

c **Identify** the distance at which the signal strength becomes **less** than the noise.

d **Describe** what would happen to the mobile phone signal at this distance, and so **explain** why it would be important to know this distance.

Telecommunications engineers use a ratio called **signal–noise ratio**. This is given by:

$$SNR = \frac{\text{signal strength (dB)}}{\text{average noise (dB)}}$$

e Use the graph to **estimate** the SNR for the cell phone at 100, 200 and 500 m distance.

f **Suggest** a way that the design of a cell phone could be changed to increase the SNR. **Explain** your answer.

Figure 7.26 is taken from an online advertisement by a company that sells special 'protective' jackets for cell phones.

g **Outline** what the thermographs (thermal photographs) claim to show.

h With reference to the Pro-TECH advertisement text, and to what you have learned about the electromagnetic spectrum, **evaluate** the claim by the company that use of a cell phone for 15 minutes could endanger the health of your brain.

■ **Figure 7.26**

Reflection

In this chapter we have **described** communication by different kinds of wave energy, and we **analysed** the oscillatory motion that produces the waves. We **explained** how different kinds of wave move through different materials, and **described** how one very important kind of wave does not seem to require any material to move through at all. We **summarized** how our personal experience of different kinds of wave energy is related to the form of the waves themselves, and we **outlined** the ways in which humans have learned to manipulate the form of waves in order to devise new and better ways to communicate. Finally, we **reflected** on the opportunities for international communication we have today, on the effects communication can have and on what **action** we can take to better communicate.

Use this table to reflect on your own learning in this chapter.		
Questions we asked	Answers we found	Any further questions now?
Factual: How do we measure a wave? What affects the speed of a wave? What is sound? What is light? What other kinds of wave are there? How can we manipulate and control light waves?		
Conceptual: How does wave motion differ from other kinds of motion? How do we experience different kinds of wave? Where is the kinetic energy in wave motion? How can a wave carry information? How does our experience of waves affect the quality of our communication? How has improved communication affected our world?		
Debatable: To what extent have improved communications made the world a 'global village'? What advantages and disadvantages might global telecommunications bring?		

Approaches to learning you used in this chapter:	Description – what new skills did you learn?	How well did you master the skills?			
		Novice	Learner	Practitioner	Expert
Critical-thinking skills					
Creative-thinking skills					
Communication skills					
Transfer skills					
Learner profile attribute(s)	Reflect on the importance of being a good communicator for our learning in this chapter.				
Communicators					

8 How is our climate changing?

Scientific **evidence** shows that human activity is leading to major **changes** in **global environments**.

MELTIN

Rai

CO_2

WATER SF

HURR

... scientists say

FLOOD WARNING

... on verge of extinction A

... END

EXI Natur

DROUGHT REFUGEES

... DISASTERS

LAST

... DESTROYED

PLANET IN TROUBLE

CLIMATE CHANGE

OIL SPILL THREATS

CATASTROPHE ENVIRON

Figure 8.1 Climate change in the news!

CONSIDER AND ANSWER THESE QUESTIONS:

Factual: Where do we obtain energy from? What are the advantages and disadvantages of different energy sources? What are renewable and non-renewable energy sources? How do we convert energy from renewable and non-renewable sources? What evidence is there that human activity is affecting Earth's climate?

Conceptual: How does energy affect matter? How does human activity affect the Earth's climate?

Debatable: To what extent is human activity responsible for climate change? To what extent can we responsibly manage Earth's environment?

Now **share and compare** your thoughts and ideas with your partner, or with the whole class.

IN THIS CHAPTER, WE WILL...

- **Find out** how the Earth's atmosphere helps maintain the conditions that make life possible
- **Explore** the physics behind the processes that keep the Earth's climate in balance, and the factors that are affecting that balance
- **Take action** locally to reduce our own impact on the global climate balance.

...EMISSIONS

Global warming crisis

Amphibian deaths ...

-CAPS

est destruction

Deforestation ...

AGE TIME FOR CHANGE

METHANE

E WARNING DESERT EXPANDING

ntists warn of

er earthquake

S WARN ...

TREES ...

aster strikes

IANCE

ars over ...

on affecting ...

ROUBLE IN ...

TAL ...

■ These Approaches to Learning (ATL) skills will be useful ...

- Critical-thinking skills
- Creative-thinking skills
- Transfer skills
- Information literacy skills
- Media literacy skills
- Communication skills
- Collaboration skills

● We will reflect on this learner profile attribute ...

- Balanced – we will reflect on our interdependence with others, and consider different perspectives on issues.

◆ Assessment opportunities in this chapter

- **Criterion A**: Defining, problem solving, categorizing, and making scientifically supported judgements
- **Criterion B**: Creating new investigations through applying concepts and knowledge already learned
- **Criterion C**: Interpreting and analysing data from laboratory models to understand real-life climate issues
- **Criterion D**: Researching and taking action on issues concerning global climate change

The Earth's climate is making news. Climate change is the biggest 'scientific story' of our times. But what is the science behind the stories? Are the news reports reliable sources of information? Is the Earth's climate changing? If so, what is causing those changes, and how quickly are they occurring? In this chapter we will be examining the scientific evidence for climate change, and we will be using physics to explore the causes and the outcomes of climate change – whether contemporary and measurable, or hypothetical for the future. Finally, we will be inquiring into the solutions that science may offer.

KEY WORDS

absorb	input
absorption	output
emit	reflect
emission	

How does energy affect matter? What affects the Earth's climate?

EARTH'S ENERGY BALANCE

The space around the Earth may look empty, but in fact it is full of energy passing through (more on this in Chapter 12). The Earth is constantly bathed in the energy that streams out from the Sun from all parts of the electromagnetic spectrum (see Chapter 7.) If the Earth had no appreciable atmosphere, it would be like the planet Mercury – a hot, hard ball of lifeless rock. Earth's atmosphere protects us from solar radiation, and life on Earth has evolved so that it is perfectly adapted to the conditions on the surface beneath this protective sunscreen.

■ **Figure 8.2** Sunscreen protects us from the small percentage of UV rays that reach the Earth's surface

The Earth's surface and its atmosphere form a complex system in which energy that passes through and energy that returns to space are balanced. Different layers in the atmosphere contain a different mix of gases and so have different properties (see Figure 8.3).

Earth's atmosphere is almost 500 km thick, but about 80% of the gas is within 16 km of the surface, in the **troposphere** – which is where all the weather occurs.

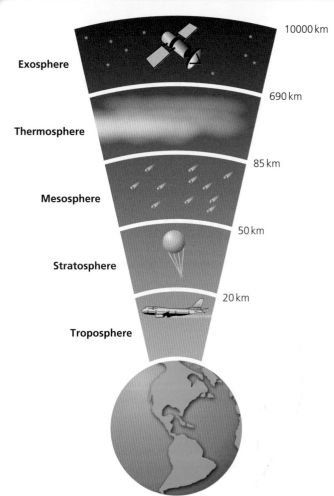

■ **Figure 8.3** Principal atmospheric layers and descriptions

While we may feel light and free while walking on a summer's day, in fact we are standing at the bottom of a sea of gases that exerts a pressure of around 1×10^5 Pascals (100 000 Newtons of force per metre squared) all over our bodies.

The energy that reaches the Earth has been **radiated** across space. Radiated energy can interact with matter in different ways, according to the wavelength of the radiation and the properties of the matter. There are three processes that can be used to characterize the interaction of radiant energy with matter.

ACTIVITY: Finding the balance

ATL

- **Critical-thinking skills:** Analyse complex concepts and projects into their constituent parts and synthesize them to create new understanding

Individually, analyse Figure 8.4 showing the way energy is absorbed, reflected, or emitted from the Earth–atmosphere system.

Global energy flows Wm⁻²

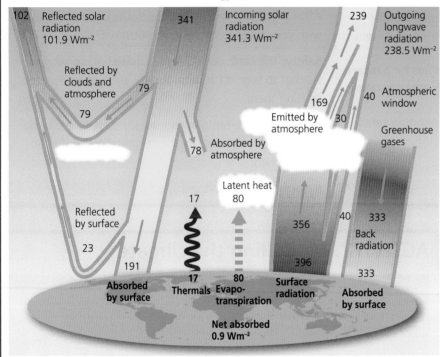

- **Figure 8.4** Inputs and outputs – climatic energy balance

In pairs, make a table with three headings: **Inputs, Outputs, Stored.**

Use your table to **categorize:**
- **sources that add to the energy content of the Earth–climate system**
- **processes that subtract or remove energy content from the Earth–climate system**
- **processes that store energy in the Earth–climate system.**

Now use the values on the diagram to **calculate** the energy balance of the Earth–atmosphere system under the conditions shown.

Comment on your calculated values.

- **Absorption** means that the radiant energy is 'soaked up' or absorbed by the atoms and molecules that comprise the material. It is converted into kinetic energy in the particles of the material, and this means that the temperature of the material increases. The energy which produces this effect is called **infra-red** energy, since it has a wavelength somewhat longer than the red region of the visible spectrum (see Chapter 7).

- **Reflection** means that the radiant energy is not absorbed but returned back on its path, without change in properties such as wavelength – a 'mirror-effect'.

- **Emission** is where matter gives out radiation. The source of this emitted energy may be energy that was previously absorbed, but on emission its properties – such as intensity and wavelength – may be changed. Energy may also be emitted due to processes that are taking place within the matter, for example, chemical or nuclear reactions.

ACTIVITY: Radiant surfaces

■ ATL

■ **Critical-thinking skills**: Interpret data; Evaluate evidence and arguments

Aim: To investigate the interaction between radiant energy and matter.

Equipment

Part 1

- **Squares of black, white and silver-coloured card**
- **Radiant-heat source (electric heater)**
- **Timer**

Part 2

- **Three metal boxes or glass beakers – one painted black on the outside surface, one painted white and the third painted silver**
- **Polystyrene (styrofoam) or cardboard lids for beakers, each with a hole for the thermometer or temperature probe**
- **Accurate thermometer or digital temperature sensor with datalogging equipment**

Hypothesis

Classify each of the three colours (white, black, silver) in terms of their effect on radiant energy. Which will be the best emitter, absorber, reflector? **Explain** your hypothesis, perhaps with reference to examples you have researched.

Research the thermal properties of different coloured surfaces using search terms: **thermal colour surfaces**.

Method

Part 1

Use a lab stand and clamps to position the three pieces of card a few centimetres from the heat source. Make sure that the cards are equal distances from the source, in a circle.

Turn on the heat source and start the timer.

After a few minutes, gently touch the **back** of the cards with your fingers. What do you observe?

Safety: The cards could become quite hot, so ask your teacher to suggest an appropriate amount of time for heating them.

Place a sensitive temperature sensor close to the backs of the cards and measure the temperature of the air behind them.

Record your observations.

The 'Activity: Radiant surfaces' shows us that different surfaces have very different effects. Of course, the Earth–atmosphere system does not have uniform black, white and silver surfaces – but there are surfaces in the Earth–atmosphere system that might behave in a similar way to these colours.
The experiment can work as a first-approximation 'modelling' of radiant energy effects.

ACTIVITY: Modelling the climate

■ ATL

■ **Information literacy skills**: Make connections between various sources of information
■ **Critical-thinking skills**: Use models and simulations to explore complex systems and issues

Look at the image of the Earth taken from space (Figure 8.5).

In pairs, **discuss** and **identify** parts of the Earth–atmosphere system that might have an effect on radiant energy.

Use a table like this one to **compare** the radiant-surfaces model to the real Earth–atmosphere system.

Part of Earth–atmosphere system	Radiant-surface model colour	How colour affects radiant energy

Part 2

Measure equal amounts of water (perhaps 100 ml) at room temperature and put the water in each of the three coloured containers. Place the lids on the containers and insert the thermometer or temperature probe.

Position the containers around the radiant-heat source in a circle at equal distances.

Turn on the radiant-heat source. Measure the temperature at suitable intervals or datalog.

Continue until you think a significant effect has occurred.

Turn off the radiant-heat source.

Continue to measure the temperature at intervals or datalog as the water cools back to room temperature.

Results

Organize and **present** your data in a way that allows you to analyse and compare the outcomes for each of the containers. Include **measurement uncertainties** as appropriate.

Analysis

Analyse your data to compare the effects of the coloured surfaces on heating, and on cooling. Try to **quantify** the effects so that you can compare with actual values, rather than qualitatively.

Conclusion

Write a conclusion about the effect of the different coloured surfaces on radiant energy with reference to your analysed values. **State** whether your hypothesis was correct, and **explain** your reasoning or make **comparisons** to similar effects with other kinds of energy.

> **Hint**
>
> We know that heating occurs due to radiated infra-red energy. This is quite close to visible light in the electromagnetic spectrum.

Evaluation

Evaluate your experiment. What evidence do you see for error? How significant were the measurement uncertainties? What might have caused them?

Suggest improvements to the experiment design that might reduce or eliminate these sources of error.

> ◆ **Assessment opportunities**
>
> This activity can be assessed using Criterion C: Processing and evaluating.

Figure 8.5 Earth from space

In this activity, we have used a simple lab experiment to 'model' the much more complex Earth–atmosphere system. While the results of the model will not tell us exactly how the input (radiant energy) affects the phenomenon (heating), they do give us a 'first approximation'. Science relies on laboratory models to understand the mechanisms that underlie complex phenomena, and science makes progress by steadily refining the accuracy of its models. The debate about climate change is often focussed on the **validity** of the models used by climate scientists.

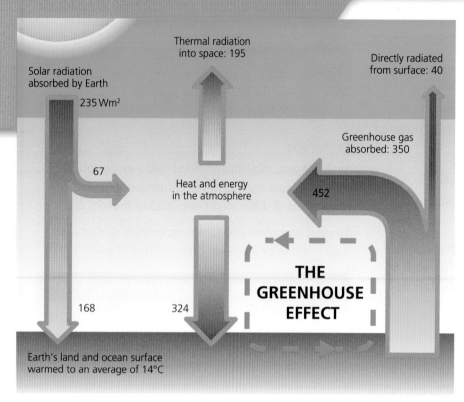

Solar radiation
absorbed by Earth
235 Wm²

Thermal radiation
into space: 195

Directly radiated
from surface: 40

Greenhouse gas
absorbed: 350

67

Heat and energy
in the atmosphere

452

THE
GREENHOUSE
EFFECT

168   324

Earth's land and ocean surface
warmed to an average of 14°C

■ **Figure 8.6** The greenhouse effect

In the atmosphere, there are no 'surfaces' as such, but gases present in the atmosphere still interact with the radiant energy of the Sun and the re-emitted energy from the Earth's surface.

The **greenhouse effect** has been identified as one of the main processes by which natural atmospheric heating occurs. We already noted that, when a surface emits radiation that it previously absorbed, the properties of the emitted radiation may be different. This is the case for the radiation that is emitted by Earth's land masses. The incident radiation is re-emitted as longer wavelength **infra-red**, which in turn can be reabsorbed and re-emitted by certain gases in the Earth's atmosphere. When the infra-red is re-emitted, some of it is directed back down towards the Earth's surface, rather than out into space. This results in the infra-red radiation becoming 'trapped' near to the Earth's surface, and so the average temperature of the troposphere increases. Without this natural heating effect, the Earth's atmosphere would probably be too cold for life.

This phenomenon is called the 'greenhouse effect' because it is somewhat similar to the way that glass panes in a greenhouse result in raised temperatures for the plants inside – although it is not actually the same effect, since in a greenhouse much of the heating is really due to reduced airflow.

ACTIVITY: Modelling the greenhouse effect

■ ATL

■ **Creative-thinking skills:** Create novel solutions to authentic problems; Apply existing knowledge to generate new ideas, products or processes

Review the 'radiant-surfaces' model experiment. **Design** a modified version of the experiment which would allow us to compare the effect of 'trapped' infra-red energy to energy that is allowed to escape from the system.

Hint

You may wish to use thin sheets of glass or transparent plastic between the heat source and the card surfaces.

◆ Assessment opportunities

In this activity you have practised skills that are assessed using Criterion B: Inquiring and designing.

ACTIVITY: Guilty gases

Warming by the greenhouse effect is caused by the presence of certain gases in the troposphere. The gases identified by climate scientists are shown in the table below.

Research using greenhouse gas properties to complete the table.

Greenhouse gas	Chemical formula	Concentration in Troposphere (ppm*)	Effect	Cause(s)
water vapour		depends on conditions		
carbon dioxide		395		
methane		0.002		
ozone		0.0003		

*ppm = **parts per million** this is the number of particles of the greenhouse gas found in every million atmospheric particles.

ACTIVITY: Where does it all come from?

Figure 8.7 shows us the main human sources of greenhouse gas in the atmosphere.

How **reliable** is the information in the chart? What do the categories chosen make you wonder? Consider what they might include, or exclude.

Figure 8.8 shows headlines from newspaper or online articles about climate change.

Analyse the information in Figure 8.7 and use it to **evaluate** the headlines in Figure 8.8. Why do they say that? What evidence is there to support the claims made in these headlines?

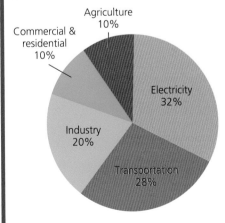

■ **Figure 8.7** Key contributions to greenhouse gases

■ **Figure 8.8** News about the climate

Where do we obtain energy from? What are the pros and cons of different energy sources?

ENERGY FOR A GLOBAL SOCIETY

While electricity is a very efficient way to transport energy over distances, the greatest contributor to human-made greenhouse gases is electricity production. This is primarily because the majority of electricity is produced using the energy stored in **fossil fuels**. Fossil fuels are so called because they are the fossilized remains of plants from a time when the Earth was much greener than it is now, known as the **carboniferous period** some 360 million years ago. The plants trapped the Sun's energy in their chemical structure in the form of long-chain carbon molecules. When we combust fossil fuels, a lot of energy is released as heat – but the process also produces large quantities of water vapour and gaseous carbon compounds, such as carbon monoxide and carbon dioxide, as well as particulate pollution in the form of dust and soot. A typical combustion equation is:

$$\underset{(g)}{CH_4} + \underset{(g)}{2O_2} \rightarrow \underset{(g)}{CO_2} + \underset{(g)}{2H_2O}$$

This example is for the combustion of methane or 'natural gas.' Note that this equation is the 'cleanest' possible combustion process, where all the carbon atoms are combined into **carbon dioxide**. Less efficient combustion leads to the production of **carbon monoxide** – which is not only a greenhouse gas but also poisonous.

ACTIVITY: Modelling greenhouse gases

■ ATL

- **Creative-thinking skills**: Create novel solutions to authentic problems; Apply existing knowledge to generate new ideas, products or processes

Water vapour, carbon dioxide and methane are all readily available in the laboratory. Use the investigation cycle to **design** an experiment to model how these gases respond to radiant-heat energy.

Safety: Before beginning your experiment, make sure that your teacher checks for safety!

◆ Assessment opportunities

This activity can be assessed using Criterion B: Inquiring and designing and Criterion C: Processing and evaluating.

Another way to compare energy sources is to use the concept of **sustainability**. How long can we continue to use these sources? There are two major limitations to the sustainable use of fossil fuels:

1 Fossil fuels will ultimately run out. The timescale for the production of fossil fuels is millions of years and insufficient new plant material is being buried to replace the fossil fuels we are using. Estimates for the lifetime of remaining fossil fuels vary widely, and new sources of fossil fuels such as shale 'fracking' are discovered. However, accessing new sources requires technical innovation and each new development brings with it new risks and new environmental impacts.

2 Fossil fuel pollution has significant environmental impact and the great majority of scientific opinion holds it to be the key cause of climate change. This means that, even if fossil fuels do not run out in the near future, their environmental impact might be unsustainable because they can cause irreversible damage to our global climate.

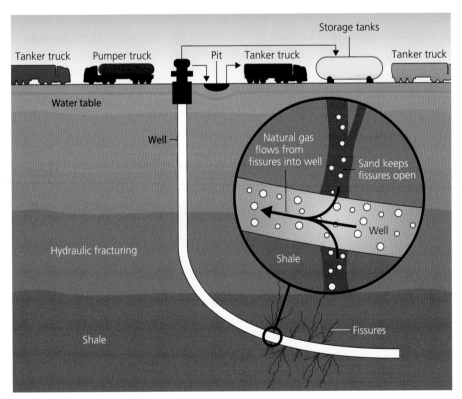

■ **Figure 8.9** New ways to exploit fossil fuels, such as shale fracking, might prolong their lifetime but bring with them new environmental concerns

ACTIVITY: Comparing energy sources

■ **ATL**

■ **Information literacy skills**: Collect and analyse data to identify solutions and make informed decisions

We might wonder why humanity chose to use fossil fuels when they produce so much pollution. In Chapter 6 we explored some of the historical reasons for this but there are scientific reasons that led us to choose fossil fuels as an energy source.

The **specific energy** of a fuel tells us now much energy is stored in a fuel per unit mass.

Research **specific energy density** to gather data on the energy densities of the different fossil fuels and non-fossil fuels in the box.

coal	natural gas
fuel oil	gasoline (petroleum)
wood	ethanol

Compare energy sources in terms of their specific energy density and write a **conclusion** about why fossil fuels have been such a popular source of energy for the last 300 years.

■ **Figure 8.10** Ethanol pump in Brazil

■ **Figure 8.11** Green biobus

In the search to replace the use of fossil fuels, a possibility is to identify new fuel sources for combustion that can be more easily replaced. There has been a lot of research and some commercial success in the use of alcohols extracted from plant matter or biomass, for example ethanol made from sugar cane or maize. This form of sustainable fuel is very important, for example, in Brazil. Equally, rapid-growing species of tree can be used to provide wood for combustion in power plants too. Since these energy sources can be replaced relatively rapidly by regrowing the plants or trees, they are more sustainable than conventional fossil fuels.

▼ Link: Geography

Grouping crops as fuel sources brings its own environmental impact in terms of land and resource use.

We already noted how the Earth–atmosphere system is alive with energy. Much of this energy is evidenced in climatic phenomena, such as wind, or directly available in the form of radiant energy from the Sun. A further possibility is that we might try to harness these energy sources. These resources will only run out when the source of their energy – the Sun – stops working, which is not expected to happen for at least 10 billion years. To all intents and purposes, they will last forever and so are sustainable. Direct energy sources and biomass sources are usually referred to as **renewable** resources because they will not run out.

Another possibility is to use the heat generated by nuclear reactions as a source of energy. We will look in more detail at how those reactions work in Chapter 11.

ACTIVITY: Fuel energy card game

As a class make a list of all the different fuel energy sources – renewable and non-renewable – that you can think of. Research using **fuel energy resources** to see if there are any you don't know about.

Categorize the energy resources as renewable, non-renewable, and sustainable.

For each of the energy sources identified, research the following factors:

- **specific energy content**
- **pollution on combustion**
- **pollution to extract the fuel**
- **pollution caused by waste products**
- **cost**
- **efficiency.**

■ ATL

■ **Transfer skills:** Combine knowledge, understanding and skills to create products or solutions

Discuss your research and **compare** the energy sources in terms of these factors, and then try to give them a 'score' from 1 to 7, where 7 means 'best' and 1 means 'worst'.

Use your scores to **design** a card game for other students who are studying climate physics. In the game, the players must try to 'trump' or beat each other by comparing the energy resources on the cards.

Have fun playing your game!

How do we convert energy from renewable and non-renewable sources?

POWER FROM NATURE

In order to produce the electricity we need, we must first capture and put to use, or 'harness', a source of energy.

When a **fuel** source is used – whether non-renewables such as fossil fuels or fuels that can be sustainably replaced – the stored energy is always released through combustion or 'burning' of the fuel. Combustion is actually a rapid chemical reaction in which oxygen combines with the fuel molecules to make new chemicals that contain less stored energy than before, and the difference in energy is released as heat – this is called an **exothermic** process.

Nuclear power is the second most important source of energy globally and many major industrialized countries rely on this form of energy production (Figure 8.12).

In a nuclear power station the stages in the energy-changing process are very similar to those in a thermal fuel power station, except that the heat is produced in a nuclear process rather than by chemical combustion. Nuclear power does not produce greenhouse gases, and some people consider that this makes it a good alternative to fuel combustion-based power production. However nuclear power is controversial for others who consider that the nature of nuclear fission, and the waste products it produces, mean that nuclear power has an environmental impact that is too high. We will look at the scientific evidence for these claims in greater detail in Chapter 11.

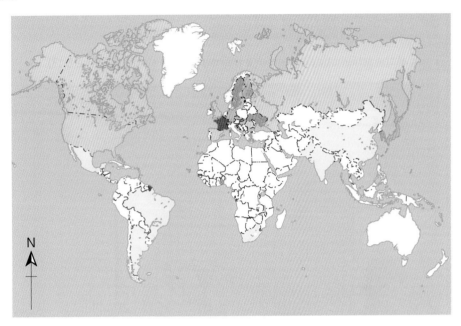

N

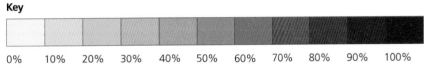

Key

| 0% | 10% | 20% | 30% | 40% | 50% | 60% | 70% | 80% | 90% | 100% |

Figure 8.12 Global map of nuclear power production

Table 8.1

Name	Input(s)	Process
hydroelectric power (HEP)	gravitational potential energy	kinetic energy (produced as water falls by gravitational force) drives turbines
tidal power	gravitational potential energy	kinetic energy (produced as sea water moves under Moon's influence) drives turbines
wave power	kinetic energy	kinetic energy of sea waves (produced by the wind) drives turbines
solar thermal power	infra-red energy directly from Sun	Sun's heat is gathered and focussed to heat water to steam to drive turbines
photovoltaic solar power	energy directly from Sun	Sun's ultra-violet radiation is used to produce electrical energy in semiconductors
geothermal power	heat energy from Earth	heat produced in the Earth's mantle is used to heat water and produce steam to drive turbines
wind power	kinetic energy	kinetic energy in wind is used to drive turbines

ACTIVITY: Burning up

ATL

- **Critical-thinking skills**: Analyse complex systems into their component parts and synthesize them to create new understanding

Look at the schematic diagram of a thermal power station (Figure 8.13).

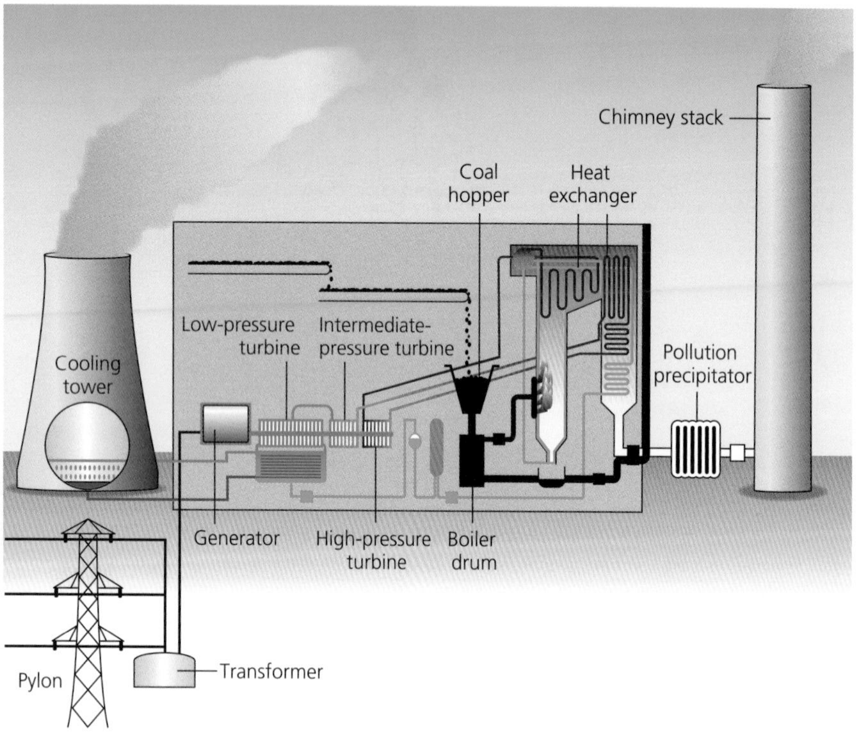

■ **Figure 8.13** Schematic diagram of a power station

Refer back to the energy-change diagrams for a steam engine and for an internal combustion engine (Chapter 6).

What were the main stages in these energy-changing processes?

The energy or fuel going into a stage can be thought of as an **input** and the energy coming out of the stage can be thought of as an **output**.

Now use the table below to **identify** the main energy changes in the power station. The first one has been done for you.

Now **analyse** Figure 8.13 to identify where energy might be wasted, and in what form it is wasted. **Outline** your thoughts in the form of a **Sankey diagram** showing the main stages in the energy-change process, and the energy wasted.

◆ **Assessment opportunities**

In this activity you have practised skills that are assessed using Criterion A: Knowing and understanding.

Input →	Process →	Output
natural gas	furnace	heat
heat		

Renewable energy resources take their energy directly from the Sun's radiated energy, from its effects on our climate, from the gravitational force of the Moon or from geological processes in the Earth. They are all used to produce electricity as the final output energy form. Table 8.1 summarizes some of these different methods of harnessing energy directly.

In fact all but **two** of these energy sources are harnessing the energy from the Sun in different forms. Which two do not?

ACTIVITY: Harnessing renewable energy

■ ATL

- **Communication skills**: Use appropriate forms of writing for different purposes and audiences; Use a variety of media to communicate with a range of audiences
- **Collaboration skills**: Listen actively to other perspectives and ideas; Build consensus

The schematic diagrams and pictures below show how energy is produced from four important renewable resources. We also looked at the operation of wind turbines as energy changers in Chapter 6.

In groups, choose one of the energy sources shown on the following double-page spread (Figures 8.14–8.21).

Research how the energy changers that harness your chosen energy source work. **Find out** about the advantages and disadvantages of each.

Outline the operation of the energy changer in the form of an energy-change flow chart and Sankey diagram showing estimated energy losses.

Prepare an information briefing in the form of a poster, computer presentation or online resource, such as a blog, that **summarizes** your research and findings.

Write a campaign speech to persuade others that your chosen energy resource is viable and should be used to contribute to our global energy needs.

In class, hold a **debate** about your research in which each group makes its campaign speech, supported by the information briefing you have made. After hearing all of the speeches, allow time for **questions** and to cross-examine each other's arguments!

Finally, try to **summarize** your class conclusions about the energy sources. What does your class recommend? Is one of these resources better than the others? Or would you advocate a mixture of them?

◆ Assessment opportunities

This activity can be assessed using Criterion D: Reflecting on the impacts of science.

➤

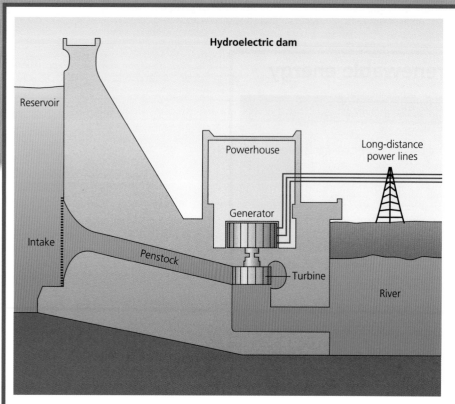

Hydroelectric dam

Reservoir

Intake

Penstock

Powerhouse

Generator

Long-distance power lines

Turbine

River

■ **Figure 8.14** Schematic diagram of a hydroelectric dam

■ **Figure 8.15** Hydroelectric power plant

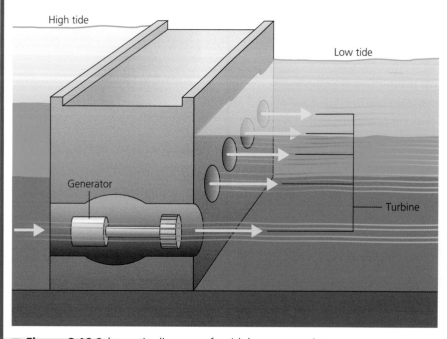

High tide

Low tide

Generator

Turbine

■ **Figure 8.16** Schematic diagram of a tidal power station

■ **Figure 8.17** Tidal power station

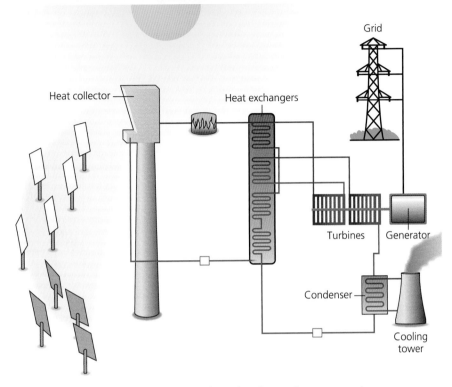

Heat collector

Heat exchangers

Grid

Turbines

Generator

Condenser

Cooling tower

Figure 8.18 Schematic diagram of a solar thermal power station

Figure 8.19 Solar thermal power station

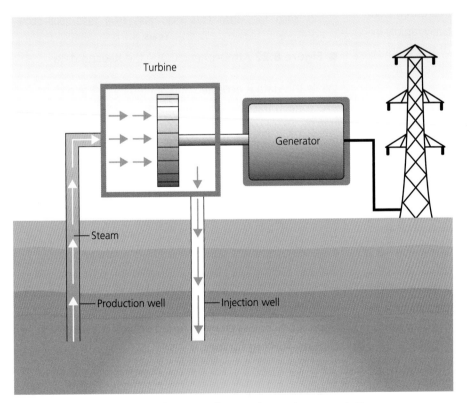

Turbine

Generator

Steam

Production well

Injection well

Figure 8.20 Schematic diagram of a geothermal power station

Figure 8.21 Geothermal power station

What evidence is there that human activity is affecting Earth's climate?

ARE THE TIMES CHANGING?

The greenhouse effect is a vital process in our atmospheric system – without it, the temperature at the Earth's surface would be around –20°C on average, rather than the current average temperature of around 15°C. This part of the greenhouse effect is the result of entirely natural phenomena, and it represents the main part of any atmospheric heating that is taking place. There is no scientific controversy about the importance of greenhouse gases on this effect, although some have disputed the relative importance of carbon dioxide compared to that of water vapour in the process. There is firm and repeatable scientific evidence that the concentration of carbon dioxide in the atmosphere has been increasing relatively rapidly in the last 200 years. The first measurements of this change were made in 1959 by the geochemist C.D. Keeling at Mauna Loa in Hawaii.

The levels of CO_2 show a rapid variation – the 'wiggle' on the graph. This corresponds to seasonal changes as plants on the Earth's surface respond to the warmer summertime. However, the overall **trend** in the data shows an increase from around 315 ppm in 1958 to 387 ppm in 2008 – that is, a 23% increase in 50 years.

Although these **direct** measurements of carbon dioxide levels began in 1958, it is possible to derive the longer-term concentrations by using trapped air from ice bubbles in the Antarctic. This extended data shows that for the last 10 000 years – since the last ice age – atmospheric carbon dioxide levels remained constant at around 280 ppm. Comparing this to the 2008 measurements gives a 38% increase.

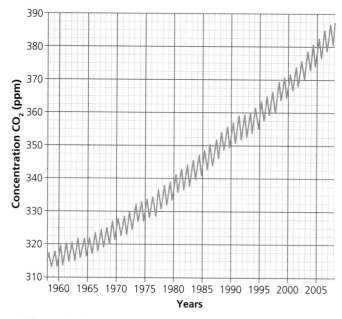

■ **Figure 8.22** Atmospheric CO_2 – the Keeling data

While this data is accepted and agreed by just about everyone in the scientific community, there has been some debate about what it means. The debate can be understood if we consider two distinct guiding questions:

1 To what extent is human activity – especially fossil-fuel combustion – responsible for this change?
2 To what extent will this result in climate change?

The idea that human activity is causing climate change, as explored by the first question, is called **anthropogenic** climate change. The alternative suggestion is that natural processes, such as volcano eruptions or changes in the amount of radiation reaching the Earth from the Sun, are responsible for this change. Together these causes are called **climate-forcing factors**.

ACTIVITY: The Sun's out

The Earth's climate definitely depends on the amount of radiation reaching us from the Sun. However, there are a number of variables that affect this quantity of radiation, even if we ignore the effects of the atmosphere discussed in this chapter. The amount of power per unit area incident on a surface is called the **irradiance** ($W\,m^{-2}$).

Look at Figure 8.23 showing the Earth in its orbit around the Sun.

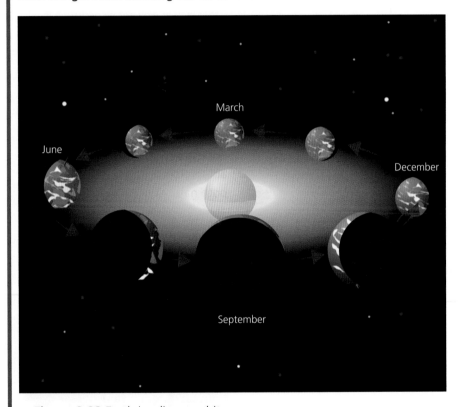

■ Figure 8.23 Earth irradiance orbit

What do you notice about the position of the Earth at the various points in its orbit?

Brainstorm variables that might affect the amount of irradiance of a certain area on the Earth's surface at a particular **latitude** as the Earth moves around the Sun.

> **Hint**
>
> What causes seasonal changes on the Earth?

From your list, **identify** a variable that could easily be controlled in a laboratory investigation.

Write a hypothesis explaining how the variable would affect the irradiance as the Earth orbits. Try to give a quantitative explanation for the change, with scientific reasoning.

◆ Assessment opportunities

In this activity you have practised skills that are assessed using Criterion B: Inquiring and designing.

The Serbian scientist Milutin Milanković (1879–1958) showed that periodic variations in the Earth's orbit and in the inclination of the Earth's axis could combine to produce significant changes in irradiance by the Sun. In turn, Milanković demonstrated that these variations could produce large-scale changes in climate on the Earth. The biggest changes occurred in a cycle of 21 000 years or so. Milanković made many of his calculations while interned as a prisoner of war during the First World War but published his results in 1916 – his predictions were not proven until climate research in the 1970s!

ACTIVITY: Global warming – evaluating the evidence

■ ATL

■ **Media literacy skills**: Compare, contrast and draw connections among (multi)media resources

Look at Figures 8.24 to 8.26.

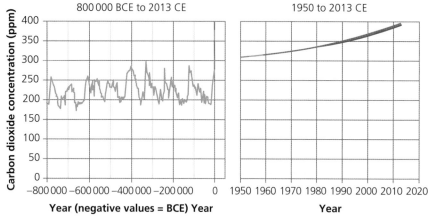

Global atmospheric concentrations of carbon dioxide over time

Data source: Compilation of 10 underlying datasets.
See www.epa.gov/climatechange/indicators/ghg/ghgconcentrations.html for specific information.

■ **Figure 8.24** Long-term CO_2 concentrations

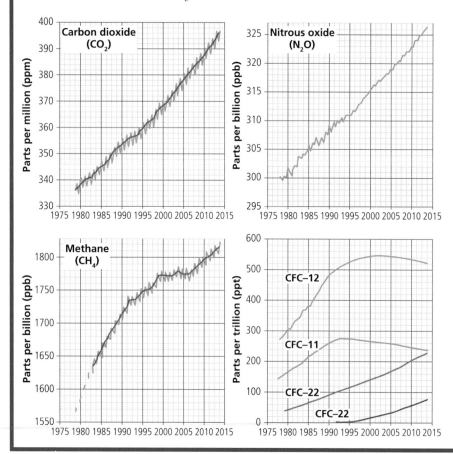

■ **Figure 8.25** Shorter term concentrations of other greenhouse gases

It is true that changes in greenhouse gas concentrations are both natural and anthropogenic. However, the human causes are insignificant when compared to the big changes that resulted, for example, in the last Ice Age. In fact, any global warming that may or may not be occurring is entirely attributable to natural causes.

■ **Figure 8.26** Extract from a blog

Think about what the data shows us about rate of change and quantity of greenhouse-gas concentrations.

Compare to what you know about global historical changes, both naturally occurring and anthropogenic.

Evaluate the blog statement with reference to quantitative data from the charts.

◆ Assessment opportunities

In this activity you have practised skills that are assessed using Criterion A: Knowing and understanding.

The second of our questions (page 150) concerns the way in which the climate actually responds to these changes in atmospheric composition – termed the **climate response**. If the response results in a change in average temperature, then climate scientists define

$$\text{climate sensitivity} = \frac{\text{change in average temperature}}{\text{climate forcing factors}}$$

To remind ourselves, there is no debate that increased concentration of greenhouse gases (whatever the cause) will result in increased atmospheric heating. The debate rather concerns the capacity for the Earth–atmosphere system to compensate for these changes.

Notice that, even if CO_2 levels have doubled over the last 200 years, this is not predicted to produce a doubling in temperature! The predictions of temperature change due to this effect alone – not including any climate-response factors – is 1.2 to 1.3°C. This is because the atmosphere will eventually reach a point of **saturation** where it is effectively opaque to re-emitted infra-red – which means that adding more CO_2 cannot increase absorption any further.

A useful concept in understanding climate response is **feedback**. Feedback means that some of the output from a system itself becomes an input – so the effect produced becomes a cause of the effect!

Feedback can be categorized as positive or negative (Figure 8.27).

When feedback is positive, it adds to the input – and causes a bigger and bigger output. This results in a system which is unstable. If feedback is negative, on the other hand, the output tends to subtract from the input – so it tends to lessen the effect and stabilize the system. When the negative feedback reaches a certain point, it will cause the system to stop changing and this is then a condition of **equilibrium**.

In the case of global warming, positive feedback mechanisms for CO_2 production would tend to increase not only the amount but also the rate at which global warming is taking place – this would be very bad news indeed. On the other hand, negative feedback in CO_2 production would tend to stabilize the system – perhaps around a new, higher equilibrium temperature.

▼ Link: Positive and negative feedback

■ ATL

■ **Transfer skills**: Make connections between subject groups and disciplines

Feedback is a very powerful concept in science, and elsewhere. **Find out** about these examples of feedback systems from other subjects:

• the effect of interest rates or inflation in economics and mathematics

• systems leading to homeostasis in biology

• amplifiers in electronics (design).

One important feedback effect is the change in reflectivity of the Earth's surface and atmosphere. The proportion of energy that is reflected from a surface is defined as the **albedo**. The polar icecaps have a very high albedo and reflect a lot of solar energy back into space, as do clouds in the atmosphere.

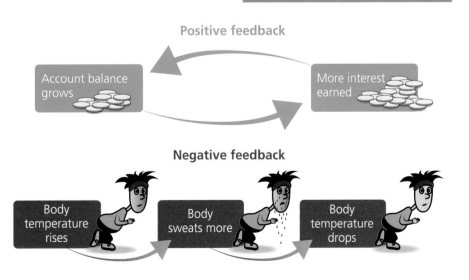

Positive feedback

Account balance grows

More interest earned

Negative feedback

Body temperature rises

Body sweats more

Body temperature drops

■ **Figure 8.27** Positive and negative feedback

COMPLEXITY, CORRELATION AND CAUSATION

One of the reasons for the controversy around climate change is that climate science – as you have seen – is very complex. It is unlikely that global warming can be attributed to any one factor alone: rather, multiple factors interact to produce the observed phenomena we have discussed in this chapter. This means that many climate predictions are based on very complex computer models which calculate the interactions of multiple factors. Some people have argued that hypotheses based on the results of these models are not reliable, because the underlying assumptions in the models may be wrong.

Since it can be hard to demonstrate the exact mechanism for climate change, it has been argued that climate scientists make their claims not from **causation**, but only from **correlation**. In other words, while we may be able to show that global warming happens at the same time as human industrial activity increased, we cannot necessarily prove that global warming was not caused by something completely different.

The Intergovernmental Panel on Climate Change (IPCC) – an international body of hundreds of professional scientists, concluded in 2014:

Human influence on the climate system is clear, and recent anthropogenic emissions of greenhouse gases are the highest in history. Recent climate changes have had widespread impacts on human and natural systems.

(Source: IPCC, Fifth Assessment Report (AR5), www.ipc.ch accessed December 2014)

If climate change is real, the Earth's climate system will certainly find a new equilibrium point. But it is unlikely that the new climate conditions will be well attuned with the needs of humanity.

What do you think?

ACTIVITY: Climate change – thinking globally, acting locally

Take action

! Tackle climate change.

Here are some ideas for taking action locally on global climate change:

1 It can be daunting, when thinking about such big issues, to find something we can do as individuals. Sometimes though, it is helpful to find out what has already been done. **Interview** older relatives – perhaps your grandparents – and ask them what they think has changed as a consequence of climate change. What examples of action do we now take for granted?

2 One of the first anthropogenic climate problems to be identified was the **ozone hole** in the upper atmosphere. **Research** about the causes and predicted effects of the hole in the ozone layer. What action was taken? What happened?

Carbon gases such as carbon dioxide are the chief drivers of anthropogenic climate change. Use carbon footprint calculator online to find a way to measure the amount of greenhouse gas generated by our daily activity.

3 **List** the chief contributors to our carbon footprint and **research** ways to reduce it. Produce an information briefing for a school assembly, town meeting or similar. **Start a club or campaign** in your school to reduce the school's carbon footprint!

You can find examples of national and international initiatives that are taking action on climate change right now – use green schools, for example, or look at the website for the Earth Day Network **www.earthday.org**

SOME SUMMATIVE PROBLEMS TO TRY

Use these problems to apply and extend your learning in this chapter. The problems are designed so that you can evaluate your learning at different levels of achievement in Criterion A: Knowledge and understanding.

THIS PROBLEM CAN BE USED TO EVALUATE YOUR LEARNING IN CRITERION A TO LEVEL 3–4

1 **Define** the following terms and **state** an example of an energy source for each.

- Renewable
- Sustainable

2 Biomass energy sources use materials that can be regrown to produce combustible fuel. The table below gives some data about different sources of biomass.

Wood type	Energy density (MJ/kg)	Annual yield tonnes per hectare
wood (willow)	13	12.9
wheat straw	13.5	3.5
biodiesel (from rapeseed)	37	1.1
bioethanol (from sugar beet)	27	4.4

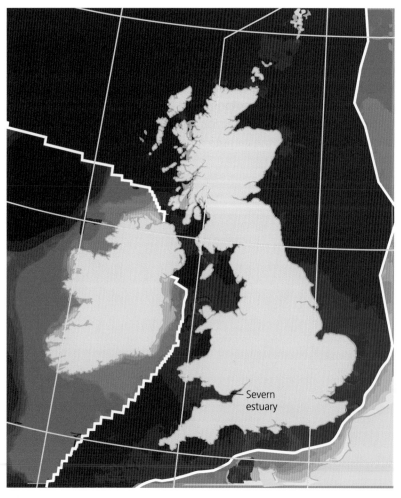

■ **Figure 8.28** Map of UK tidal patterns showing river Severn

Severn estuary

Key

Power (kW/m²)

- 20.01–50.00
- 10.01–20.00
- 8.01–10.00
- 6.01–2.01
- 0.76–1.00
- 0.51–0.75
- 0.26–0.50
- 0.06–0.25
- 0.01–0.05
- 0.00
- Land
- UK Continental shelf & Channel Island territorial sea limit

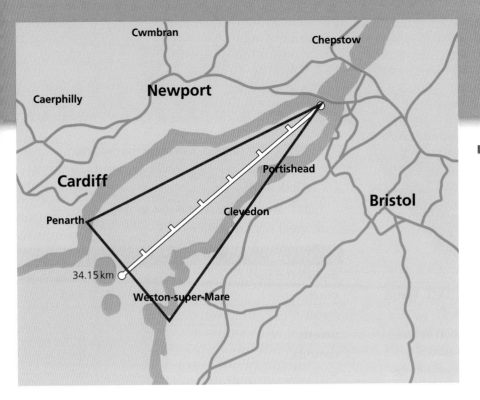

Figure 8.29 Aerial map of the river, showing barrage and length

Jose is a farmer who wants to convert his farmland to biomass-fuel production. He has a small farm of 50 hectares.

a If Jose chooses to plant his land with each of the fuels in the table, use the data to **calculate** the theoretical energy available per year from each of the fuels.

The different biomass fuels require different kinds of energy-changing process. The table below shows the efficiencies of the different generators used to produce electricity from each of the fuels.

Fuel type	Combustion efficiency (%)
wood pellets/chips	80
biodiesel	50
bioethanol	30

b Use the data on efficiencies to **calculate** the energy output from each of the fuel options.

c **Compare** your results and **interpret** them to decide which biomass fuel would be the best to use.

d **Outline** why energy production through the combustion of biomass fuels may be renewable, but not environmentally sustainable.

THIS PROBLEM CAN BE USED TO EVALUATE YOUR LEARNING IN CRITERION A TO LEVEL 7–8

3 One way to harness the energy of the tides is to use barrages built across rivers which experience particularly high tidal flows.

a **Describe** how a tidal barrage generates electricity.

In the 1990s a proposal was made to use the estuary of the river Severn in the United Kingdom as a tidal power source (see Figure 8.28).

One proposal is to build a barrage across a part of the river that is 16 km wide, between Penarth and Weston-super-Mare. The tidal range at this point is 14 m (see Figure 8.29).

b **Estimate** the area of the Severn estuary as shown by the red triangle.

c **Calculate** the volume of water entering the estuary at high tide, assuming that the water rises by 14 m.

d **State** the average value for the change in depth of the water due to the high tide.

e Assuming the density of seawater = 1000 kg m^{-3}, **calculate** the change in gravitational potential energy in the tidal seawater that would be stored behind such a barrage.

f There are two high tides every day. **Show** that the power available from one of the Severn tides is approximately 12 GW.

g Calculations of the likely realistic energy yield from the barrage gave around 8500 MW. **Estimate** the efficiency of the barrage.

h One proposal claimed 'a tidal barrage across the river Severn could provide between 5–15% of the UK's national energy needs'. Carry out your own **research** and make suitable **estimations** to **evaluate** this claim.

4 Leticia and Marco carry out an investigation to find
the effect of changing the colour of the surface of
some containers on the temperature of water in the
containers. They paint five containers with different
mixtures of white and black paint, then heat the
containers with a heat source. The containers are placed
at equal distances from the heat source.
The table below gives their results.

Percentage of black paint	Temperature change after 5 minutes (°C)
0	3.75
20	12.75
40	18
60	21
80	24

a **Plot** a suitable graph to show their results.

Leticia thinks that the graph shows a linear relationship
between temperature change and percentage black
paint content. Marco thinks that it does not, and does
some research. He finds out that the temperature is
proportional to the square root of the **albedo**.

b **Plot** a new graph that will allow you to test Marco's
claim. **State** whether you think Marco's or Leticia's
claim is best supported by the data.

c Using Marco's graph, **estimate** the temperature of a
container that is painted with 50% black paint.

d Using Marco's graph, **suggest** a possible formula
for the relationship between the percentage of black
paint in the mixture and the temperature increase of
the water in the containers.

e Using your formula, **calculate** the predicted
temperature rise for a container coated with 100%
black paint.

The table below shows the albedo values for some
different kinds of surface.

Surface type	Typical albedo
asphalt (tarmac)	0.04
ocean	0.10
forest	0.15
desert	0.40
concrete	0.55
snow	0.80

f Using the data in the table, **interpret** the likely
effect on average albedo of:
- increasing urbanization
- melting polar caps
- deforestation.

For each of the above, **outline** whether you
think it will result in a positive, a negative or a neutral
feedback response in average temperatures.

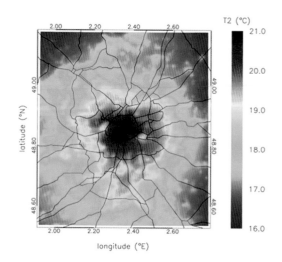

■ **Figure 8.30** Paris heat map

Figure 8.30 shows a thermograph of mean temperatures
for one day in summer 2003 in Paris, France.

g With reference to Leticia and Marco's findings, and
to the albedo data above, **interpret** the image to
explain what might be meant by the phrase 'urban
heat island'.

Reflection

In this chapter we have explored the physics of the Earth's complex climate balance. We have examined and evaluated the evidence that rapid change is occurring in that system, and we have compared the effects of climate change, considering whether they may result in positive feedback leading to instability, or negative feedback tending to equilibrium. We have explored the reasons for the debate and controversy about climate change through examination of the scientific claims made by both sides.

If we are convinced by the scientific evidence that anthropogenic climate change is real and happening, it is important that we do something about it. After all, the responsibility for these changes lies with us.

Use this table to reflect on your own learning in this chapter.		
Questions we asked	Answers we found	Any further questions now?
Factual: Where do we **obtain** energy from? What are the advantages and disadvantages of different energy sources? What are renewable and non-renewable energy sources? How do we convert energy from renewable and non-renewable sources? What evidence is there that human activity is affecting Earth's climate?		
Conceptual: How does energy affect matter? How does human activity affect the Earth's climate?		
Debatable: To what extent is human activity responsible for climate change? To what extent can we responsibly manage Earth's environment?		

Approaches to learning you used in this chapter:	Description – what new skills did you learn?	How well did you master the skills?			
		Novice	Learner	Practitioner	Expert
Critical-thinking skills					
Creative-thinking skills					
Transfer skills					
Information literacy skills					
Media literacy skills					
Communication skills					
Collaboration skills					
Learner profile attribute(s)	Reflect on the importance of being balanced for our learning in this chapter.				
Balanced					

9 Are all our futures electric?

The ***development*** of electrical ***systems*** has defined the modern world and made ***new futures*** possible.

■ **Figure 9.1** Aurora

CONSIDER AND ANSWER THESE QUESTIONS:

Factual: How do we gain energy from electricity? How do we control electricity? How do we measure electricity? How much does electricity cost?

Conceptual: How do electrical systems help us develop new ways of living? How can we use electricity safely?

Debatable: What is the balance between benefits and costs of electricity? Is it realistic to think we can reduce the costs?

Now **share and compare** your thoughts and ideas with your partner, or with the whole class.

IN THIS CHAPTER, WE WILL...

■ **Find out** how electricity is used to do useful work, how it is controlled, how it is measured, how we can use it safely and how much it costs.

■ **Explore** the ways in which materials conduct electricity, different kinds of circuit, and how to build safety systems into circuit design.

■ **Take action** to measure the cost of electricity and how we can reduce our electricity consumption, our electricity bills, and so reduce our impact on the environment.

■ These Approaches to Learning (ATL) skills will be useful ...

■ Information literacy skills

■ Communication skills

■ Critical-thinking skills

■ Creative-thinking skills

■ Collaboration skills

■ Reflection skills

■ Affective skills

■ Transfer skills

■ **Figure 9.2** Lightning

● We will reflect on this learner profile attribute…

● Caring – we will consider the importance of safety in using electricity and also the importance of limiting our energy consumption in the interests of the environment.

◆ Assessment opportunities in this chapter

◆ **Criterion A**: Knowing and understanding
◆ **Criterion B**: Inquiring and designing
◆ **Criterion C**: Processing and evaluating
◆ **Criterion D**: Reflecting on the impacts of science

KEY WORDS

charge	current	power
conductor	insulator	resistance
conductivity	potential	

Nature is electric. In Chapter 2 we saw how electric fields help to hold everything together at the atomic scale. Electricity is present in the visible natural world, too – lightning strikes are its most visible phenomenon, but the Earth is surrounded by electric fields that produce other incredible light displays such as the *aurora borealis* or *aurora australis* (northern and southern lights).

The power of electricity is evident to anyone who has seen the result of a direct lightning strike. However, it wasn't always obvious that lightning was electricity, and it was even less obvious how to harness this energy and put it to good use.

One of the most famous such attempts was that reported by Benjamin Franklin in 1750. Franklin had proposed earlier in that year that electricity might be gathered from storm clouds, and he had carried out some experiments using iron rods attached to the chimneys of his house.

■ **Figure 9.3** Benjamin Franklin and his kite experiment

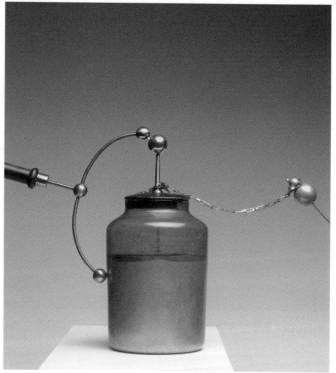

■ **Figure 9.4** A Leyden jar

It is unlikely that Franklin actually did fly a kite in a storm and wait for it to be hit by lightning, as is often thought. Rather he probably flew the kite in what looked like promising storm clouds – having first suitably insulated himself – and then noted how a key attached to the wet string of the kite emitted sparks when another (grounded) object was brought close to it (Figure 9.3). In fact, Franklin claimed to have charged a Leyden jar (Figure 9.4) this way, which was an early kind of capacitor or charge-storage device.

It was fashionable in the polite (wealthy) society of 18th and 19th-century Europe to use electricity in parlour tricks or shows for one's friends. But alongside the fun, serious science was being done. A number of ways for generating static electrical charges were devised, all of them using the principle of frictional charging that we saw in Chapter 2.

ACTIVITY: Electrostatic generators

■ ATL

- **Information literacy skills**: Access information to be informed and inform others
- **Communications skills**: Take effective notes in class

Use **electrostatic generator**. Research **two** different devices and make notes using these guiding questions:
- **How does the device generate electrical charge?**
- **How does the device store the electrical charge?**
- **How much charge, energy or potential difference can the device generate?**

The Van der Graaff generator was invented in 1929 and is one great way to see how static electrical charges behave.

Figure 9.6 shows a student touching a fully charged Van der Graaff generator – with a shocking effect on his hairstyle!

■ **Figure 9.5** Van der Graaff generator

ACTIVITY: Shocking physics

■ ATL

- **Critical-thinking skills**: Interpret data; Recognize and evaluate propositions

Physics teachers are fond of alarming their students by quoting the voltage (potential difference) carried by the dome of a Van der Graaff generator. However, it is usually perfectly safe to touch the dome, or even allow it to discharge through your body – provided some safety precautions are taken.

Table 9.1 below compares the electricity produced when a Van de Graaff generator discharges to that conducted from a standard household mains supply in a country which uses 220 V as a supply.

■ **Table 9.1**

	Van de Graaff generator	Household mains supply (highest load)
typical potential difference (volts)	100 000	220
typical current on discharge (amps)	2×10^{-3}	13
typical time of discharge (seconds)	1×10^{-3}	Until fuse blows, approx. 0.1 s
typical energy of discharge (joules)	0.5	286

Compare the values for the Van der Graaff generator and the mains supply.

What makes you think that it might be relatively safe to touch the van der Graaff, but absolutely not safe to touch the contacts of a mains supply socket?

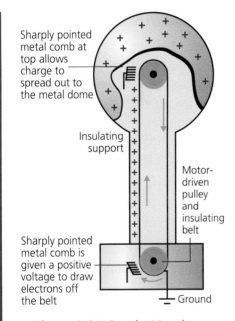

Sharply pointed metal comb at top allows charge to spread out to the metal dome

Insulating support

Motor-driven pulley and insulating belt

Sharply pointed metal comb is given a positive voltage to draw electrons off the belt

Ground

■ **Figure 9.6** Using the Van der Graaff generator

How do we gain energy from electricity? How do we measure electricity?

GOING WITH THE FLOW

Electricity is the result of electrons moving in order to remove a difference in electric charge between two points. Of course, when the electrons move, they carry kinetic energy and this energy can be used to do useful work.

In our café analogy, the waiters are like the electrons, carrying energy around the café to give to the customers. The waiters leave one side of the bar with a full jug of 'energy' and are pulled around the café to the other side of the bar, emptying the energy on the way. Of course, the amount of energy they must carry will depend on the number of customers demanding it, and how thirsty they are!

The flow of energy carried by electrons is called an electric **current**. The paths the waiters follow around the café are like the wires in an electric **circuit**.

The amount of electricity flowing in a circuit is measured as electric current in **amperes**, or **amps** for short.

We can represent circuits by drawing them as they actually look, but this tends to be unreliable and may be misleading. Consequently, scientists and engineers reduce circuits to their most basic schematic form as a **circuit diagram**.

We can measure the electric current flowing in a circuit using an ammeter. The symbol used for an ammeter in circuit diagrams is shown in Figure 9.9.

An ammeter 'samples' the electric current passing through it, without using a measurable amount of that current. For this reason, an ammeter has to be part of the circuit, so that the current can flow through it.

To make the current flow, we need a source of electrical energy. This can be a battery (or 'cell'), or a power supply unit (P.S.U.).

ACTIVITY: The electric café

■ ATL

- **Critical-thinking skills**: Use models and simulations to explore complex systems and issues
- **Creative-thinking skills**: Generate metaphors and analogues
- **Reflection skills**: Identify strengths and weaknesses of personal learning strategies (self-assessment); Consider content

■ **Figure 9.7** A café

The café in Figure 9.7 is the Electric Energy Café. In the café, waiters carry jugs of special energy drink to the customers at the tables. The waiters can only travel between the tables, where the way is clear. Some of the customers are thirstier for energy drink than others, and take more from the waiters' jugs. When the jugs are empty, the waiters have to return to the bar to refill them. If the café is very busy, then more waiters are required to keep the customers' glasses full (Figure 9.8).

Set up an imaginary electric café in your classroom. Choose waiters and customers and a person to run the bar, and then set the café in motion!

Figure 9.9 The ammeter symbol

Figure 9.8 Schematic diagram of waiters circulating in the café

Discuss your simulation as a class. Read the following paragraph, and then **consider** the following questions:

- **What did I learn from the simulation?**
- **What don't I understand about the simulation?**
- **What questions do I have now?**

Record your reflections about these questions.

When you have finished this chapter, reread your ideas about the simulation.

- **Are there any limitations to the simulation that you did not recognize before?**
- **What new questions do you have now?**

The electric café is an analogy for the way electricity can be used to carry energy. We know already that static electric fields exist where there are differences in electric **charge** and that negative charge is carried by electrons (Chapter 2). We also know that differences in energy or **potential** tend to get evened out and, where this leads to the movement of matter, it produces **kinetic** energy.

ACTIVITY: Going with the current

ATL

- **Critical-thinking skills**: Evaluate evidence and arguments; Draw reasonable conclusions and generalizations
- **Collaboration skills**: Listen actively to other perspectives and ideas; Build consensus

How does current flow in a circuit? In this experiment you will measure the amount of current flowing at different points in two different kinds of circuit.

Inquiry: To measure the electric current at different places in a circuit.

Method

Build the circuits as shown in Figure 9.10 and 9.11. Your teacher will provide the appropriate equipment. The circuits are shown as pictures, but also as circuit diagrams.

Hypothesis

Will the current be different in different parts of the circuit, or the same?

Explain what makes you think this.

Make sure that you connect the energy source and the ammeters the correct way around – the ammeters should not read less than zero amps!

Press the on–off switches and observe what happens to the bulbs. Note down your observations.

Place an ammeter in the circuit next to each of the switches. (Each time you should disconnect the switch, insert the ammeter, and then reconnect the circuit.)

Measure the current at each switch using the ammeter.

➤

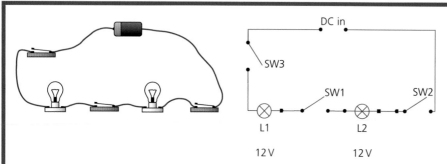

■ **Figure 9.10** Circuit 1 picture and circuit diagram

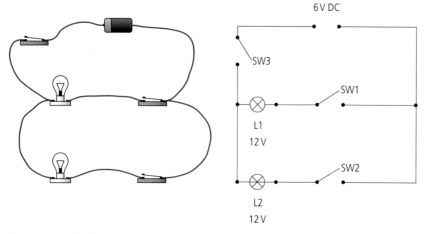

■ **Figure 9.11** Circuit 2 picture and circuit diagram

Results

Organize your observations of the effects of switches and measurements of electric current in a suitable table.

Share your data with other groups and **compare**.

Conclusion

Can you see any patterns? What is happening to the flow of electric current in each circuit? **Evaluate** the hypothesis you made, now that you have results.

The first kind of circuit is called a **series circuit** because the devices or 'loads' in the circuit come one after another in a line, or series. The second kind of circuit is called a **parallel circuit** because the circuit branches and the loads are placed beside each other, in parallel.

Discuss your conclusions, first as a small group (or with your partner) and then present to the class. **Listen** to the ideas of other groups in the class and then agree on a class **summary** of the properties for each type of circuit, series and parallel, and write it down.

◆ **Assessment opportunities**

In this activity you have practised skills that are assessed using Criterion C: Processing and evaluating.

Physics for the IB MYP 4&5: *by Concept*

Did you notice that it was important to connect the electrical energy source and the ammeters the correct way around? This is because we are using **direct current** or **dc**. Direct current always flows in one direction around the circuit.

The source of energy in the circuit – a battery, or power supply unit – causes energy to flow in the circuit by introducing a difference in charge across its two terminals. One pole has a relatively positive charge (+) and is usually coloured red, and the other has a relatively negative charge (−) and is coloured black (though, of course, the colour of the wire coating makes no difference to the electrons inside – to them, it's just a wire!)

We know the electrons carry negative charges, so naturally we would expect the electrons to be repelled away from the negative terminal and towards the positive terminal. This flow of electrons is called **real current**. However, for reasons as much historical as scientific, circuits are often analysed in terms of **conventional current** which is held to flow from positive to negative – so the opposite direction to the flow of electrons!

The current I in amperes is then given by

$$I = \frac{q}{t}$$

where q is the electric charge in coulombs and t is the time taken for it to flow in seconds, so

1 amp = $1\,C\,s^{-1}$

Battery life is often measured in **amp-hours (Ah)**. A battery with a charge life of 1 amp-hour will provide a current of 1 amp for 1 hour before it is completely discharged.

ACTIVITY: The right circuit

■ ATL

- **Critical-thinking skills**: Test generalizations and conclusions

Use the data and observations from your experiment to answer the following questions.

State which circuit type would be best to use for street lighting: series or parallel? **Justify** your answer with reference to your observations.

Compare the current flowing in each circuit, 1 and 2 (Figures 9.10 and 9.11). **State** which circuit drew the most current from the energy source.

Calculate the amount of charge that would flow in each circuit if it was left running for 5 minutes.

Suggest which circuit would run the longest on battery power and **justify** your answer.

◆ Assessment opportunities

In this activity you have practised skills that are assessed using Criterion A: Knowing and understanding.

How do we control electricity?

ACTIVITY: Superconductors

■ ATL

- ■ **Information literacy skills**: Access information to be informed and inform others
- ■ **Communications skills**: Take effective notes in class

Search YouTube for superconducting materials and then high temperature superconductors.

Take notes to help you answer these guiding questions:

- ● **What is 'superconduction'?**
- ● **Under what conditions is it achieved?**
- ● **What are the limitations of this technology at the moment?**

CONTROLLING ELECTRICITY

If you were changing a light bulb in your home, which of the Figures 9.12 to 9.13 would be advisable? Why do you think this?

■ **Figure 9.12**
Wellington boots

■ **Figure 9.14** High-tension insulator

If you look carefully at high-voltage electricity supply cables, you will notice that the cables are suspended using devices that are, in fact, made from a kind of glass or ceramic material (Figure 9.14).

However, recent developments in superconducting materials have found other ceramic materials that conduct electricity almost perfectly. In Chapter 1 we encountered graphene, a super-thin and super-strong, conductive form of graphite.

Certain materials conduct electricity better than others. Since electrons carry electrical energy, those materials which allow the greatest electron mobility are the best conductors.

■ **Figure 9.13** Wet feet

Table 9.2 below compares the structures and electrical properties of three common materials.

▮ Table 9.2

	Structure	Electron bonding	Conductivity (siemens per metre, $S\,m^{-1}$)	Conductive type
copper	metallic crystal	metallic – mobile electrons	6×10^7	conductor
carbon (as graphite)	planar crystal	covalent – with a delocalized electron	2×10^5	semiconductor
sulfur	crystal	covalent – shared, fixed electrons	5×10^{-16}	insulator

▼ Link: Chemistry
You may have encountered different bonding types through study of Chemistry.

Covalent bonds involve the sharing of one or more pairs of electrons between two adjacent atoms, so that the shared electrons are bound into molecular energy orbitals of the atoms they are bonding together – so the electrons can't move from atom to atom.

Metallic materials are characterized by bonds between atoms which 'share' many electrons across many, many atoms – thousands even. This means that electrons can move relatively freely in these materials and so they can carry kinetic energy.

Finally, there are a number of elements that fall between the two types. These materials have electron structures that leave one or two electrons with some freedom to move. In the case of carbon in the graphite form, one electron is 'free' from the crystal structure and this electron is effectively delocalized like those in a metal. These substances do conduct, but not very well and not in all circumstances (in fact, they tend to conduct **better** when they contain other atoms as impurities), and they are called **semiconductors**. This last type of material is particularly important in modern electronics.

Of course, air would be an example of a mixture of gases that contains no conducting elements. Does air conduct? One would hope not – otherwise, presumably, electric current would be leaking out from the sockets in our homes all the time and giving us unpleasant electric shocks. On the other hand, the very first image in this chapter is of a lightning strike – and perhaps you have been able to observe electric discharge through the air from a Van der Graaff generator or other charge accumulator. Are you sure air does not conduct?

The answer is that it might, sometimes, if the conditions are right. Specifically, if the charge difference between two points is great enough, then the electric field between those points will contain enough energy to cause electrons to move – and an electric current will flow as a discharge, or spark.

Electrons | Metal atoms | Sea of delocalized electrons

▮ Figure 9.15 Metal bonding with delocalized electrons

The difference in electrical energy caused by a charge difference between any two points is called the **potential difference (p.d.)**. Since it is measured in **volts (V)**, the p.d. is often called the 'voltage'.

$$\text{potential difference (V)} = \frac{\text{energy difference (J)}}{\text{charge (C)}}$$

So

$$1\,V = 1\,J\,C^{-1}$$

■ **Table 9.3** Some typical voltages

Example	Typical voltage (V)
voltage in a nerve cell in your finger	75×10^{-3}
small single cell or battery (rechargeable)	1.2
small single cell or battery (non-rechargeable)	1.5
average mains voltage in Japan	100
average mains voltage in North America	120
average mains voltage in Europe	220
average voltage of rail for metro or public transportation train	750
(long-distance) electric power lines	110×10^{3}
lightning	100×10^{6}

We can measure the potential difference between two points in a circuit using a **voltmeter**. Because the voltmeter has to 'sample' the potential at two different points, it is always connected in parallel with the circuit under test. The circuit symbol for a voltmeter is shown in Figure 9.16.

■ **Figure 9.16** Voltmeter symbol

The higher the potential difference, the greater the electric current that will attempt to flow in order to remove the difference in energy (and the greater the electric charge that will be carried between the two points concerned – see Figure 9.17). Of course, if the material between the two points was a perfect conductor, the current flow would be instantaneous and infinitely large. However, there are no perfect conductors … as far as we know.

Another analogy to help us understand the way electric current behaves is to compare it to water. In Figure 9.17, water is being pumped around a circuit made from tubing by a small electric motor. You can see that the pipe 'forks' at the top – this is like the junction in the parallel circuit in the 'Activity: Going with the current' (page 165). The pipe also has clamps attached which make it narrower.

The funnel arrangement is a current meter (like an ammeter) because the level of water in the funnel will depend on the amount of water flowing in and out of the funnel over a given time.

If we increase the speed of the electric pump, the pressure on the water will be greater.

If we reduce the diameter of the pipes using the clamps, this will make it more difficult for the water to flow.

What will happen to the level of the water in the current meter in each of these cases?

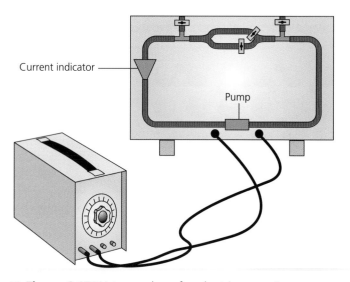

Current indicator

Pump

■ **Figure 9.17** Water analogy for electric current

ACTIVITY: Realizing the potential

What determines the amount of electric current flowing through a conductor? We already know that a potential difference across the conductor is required to make the electrons move and carry energy. But what is the effect of the conductor material?

■ **Figure 9.18** Industrial and domestic cables

Inquiry: To investigate the effect of conducting material on the electric current.

Equipment
- **Voltmeter**
- **Ammeter**
- **Connecting wires and two connecting clips**
- **Power supply unit or battery pack**
- **Variable resistor (rheostat)**
- **Various pieces of conducting wire (your teacher will supply these)**

Method

Connect the circuit components as shown in Figure 9.19.

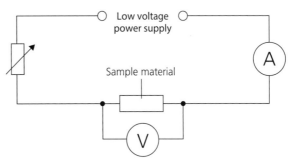

■ **Figure 9.19** Circuit diagram

Connect one of the conducting materials between the connecting clips in the circuit as shown.

Variables

Discuss: What independent variable will you change in this experiment? How will you change it? How will you measure its value?

What dependent variable will you measure?

What must be controlled (kept the same)?

Hypothesis

Suggest a relationship between the independent and the dependent variables you have identified. **Justify** this hypothesis in terms of what you know about electric current in circuits.

Before you begin, carry out preliminary measurements with a sample of material and decide on the appropriate range and interval for your readings.

Look carefully at the meters with which you have been provided. What is the smallest measurement they can make? This is called the minimum scale division. Write down this smallest scale value.

➤

Results

Make suitable measurements of the way the current in the circuit is affected for different materials. Record these measurements in a table.

At the top of the table, record the minimum scale uncertainty. For example, if your voltmeter has a scale that looks like Figure 9.20, the smallest scale division is 0.1 V. The minimum scale uncertainty is usually taken to be half this value, because you can more or less estimate whether the needle of the meter is above or below half way between the scale divisions.

■ **Figure 9.20** Diagram of voltmeter scale and needle

The minimum scale uncertainty is the limit of accuracy of your meter, such that when you read the meter you could be 'wrong' by this amount. You would then write the minimum scale uncertainty for your voltage values as

potential difference (Volts) ± 0.05 V

Analysis

Present the data in a suitable way so that you can find a relationship between the potential difference, V, and the current, I.

Conclusion

Identify a relationship in your data. **State** the relationship.

Now **share** your results with the class and **compare** to the results for other groups.

What was similar about the results? What was different about the results?

Evaluate the hypothesis you made, now that you have results.

Evaluation

Review your results. **Comment** on the following:

- **How consistent are they? Do they all fall close to the expected relationship, or are they 'scattered' around it?**
- **Was there an intercept on your graph? Is this to be expected, or is it an error?**
- **How significant (important) was the minimum scale uncertainty to the error in your results? Were there other sources of error?**
- **What could you do to minimize these sources of error?**

◆ Assessment opportunities

In this activity you have practised skills that are assessed using Criterion C: Processing and evaluating.

When a material exhibits a linear relationship between p.d. (V) and current (I), it is said to be **ohmic** (after the German physicist, Georg Ohm). Not all materials are ohmic, but metals are under usual conditions. For ohmic conductors the relationship is directly proportional, but the amount of current flowing for each volt of p.d. will depend on the conductive properties of the material. The proportionality is called the **resistance** (R) of the material, where

$$V = IR$$

and R is measured in **ohms**, which have the Greek symbol omega, Ω. This relationship is called **Ohm's law**.

Table 9.4 shows the **resistivity** of some common materials.

■ **Table 9.4** Typical resistivity of common materials

Material	Typical/average resistivity (ohm m)
gold	2.21×10^{-8}
copper	1.71×10^{-8}
iron	9.71×10^{-8}
silicon	30.0
glass (pyrex)	4×10^5
rubber	1.00×10^{15}
water (distilled)	1.80×10^5
water (fresh, tap water)	2.00×10^2
water (salty seawater)	2.00×10^{-1}

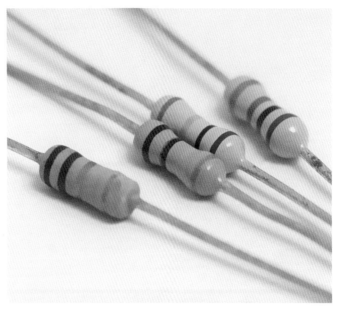

■ **Figure 9.21** A commercial resistor

■ **Figure 9.22** Resistor circuit symbol

Resistivity is the inverse of conductivity, so

$$\text{resistivity, } \rho = \frac{1}{\text{conductivity, } \sigma}$$

Resistivity is the resistance of a material for a unit area per unit length – in other words, for a fixed size of material.

Resistance causes energy to be lost in a conductor, in the form of heat. This is why resistance results in a potential difference. Resistance can be a problem, as we will find out in the next section – however, it can also be a very useful thing. Carefully controlling resistance in a circuit enables us to control the p.d. and current, too. Circuit components called **resistors** are precisely engineered to a known resistance, with a certain percentage uncertainty.

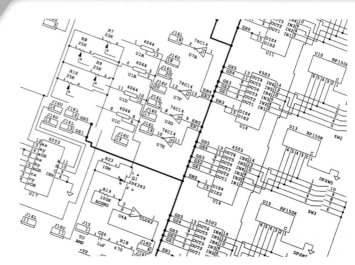

ACTIVITY: Resist, resist!

■ **ATL**

■ **Affective skills:** Demonstrate persistence and perseverance

How do resistors help us to control circuit p.d. and current?

We have already explored **series** and **parallel circuits**. How do resistors behave in these combinations?

Inquiry: To measure the effect of resistors in circuit combinations.

Method

Build the circuits in Figures 9.24 and 9.25 using fixed resistors.

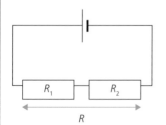

■ **Figure 9.24** Series resistors

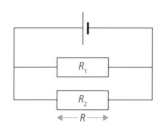

■ **Figure 9.25** Parallel resistors

Hypothesis

What do you expect to be the total resistance of the combinations shown as *R* here? Which resistor network will result in the highest total resistance? **State** your hypothesis for each resistor network and then **justify** with scientific reasoning about currents, potentials and resistances.

Now use a resistance meter to measure the total resistance of each combination.

Conclusion

Was your hypothesis correct?

If not, read on and then return to your explanation to **reflect** on what you misunderstood.

After you have learned more, write a new conclusion that improves on the explanation you gave for your hypothesis.

◆ Assessment opportunities

In this activity you have practised skills that are assessed using Criterion B: Inquiring and designing.

We can understand how resistor networks behave by analysing the way p.d. and current work in the circuits.

When in series, we need to remember that each resistor causes energy to be lost. The energy carried by the electric current is measured as p.d., but the total amount of current flowing remains the same because we don't actually lose any electrons in the resistor. This is rather like saying the customers in the electric café (page 164) don't eat the waiters, they just consume the energy drink the waiters are carrying. Alternatively, in the water analogy, the resistance of the clamped tubes causes the water to flow more slowly, but the total amount of water does not change.

So the resistors in series each cause a p.d., but the current through all resistors is the same (Figure 9.26).

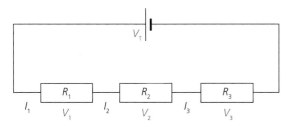

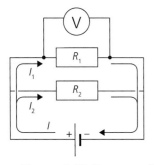

■ **Figure 9.26** Potential difference summing over series resistors

■ **Figure 9.27** Current splitting across parallel resistors connected between points X and Y

The total p.d. over all the resistors in series is given by

$$V = V_1 + V_2 + V_3 + \ldots$$

and so on. Since

$$V = IR$$

then

$$IR = IR_1 + IR_2 + IR_3 + \ldots$$

All the currents, I, are the same here, so they can be cancelled to leave

$$R = R_1 + R_2 + R_3 \ldots + R_n$$

In the parallel circuit (Figure 9.27), on the other hand, the current is splitting up to flow through each of the resistors in the network, so

$$I = I_1 + I_2 + I_3 + \ldots$$

However the p.d. across **each** resistor must be the same, since they are all connected between the same two points X and Y. From Ohm's law,

$$I = \frac{V}{R} = \frac{V}{R_1} + \frac{V}{R_2} + \frac{V}{R_3} \ldots$$

Since V is the same across all resistors, it again cancels to give

$$\frac{1}{R} = \frac{1}{R_1} + \frac{1}{R_2} + \frac{1}{R_3} \ldots + \frac{1}{R_n}$$

How can we use electricity safely?

ELECTRICALLY SAFE

Look at the images in Figures 9.28 and 9.29. Why are the birds safe to warm their feet on the high-voltage cables, while the child flying the kite is in grave danger?

We saw earlier in this chapter that the Van de Graaff generator can be safe to use in school, even though it produces very high voltages, while the discharge from a mains socket at much lower voltage can be lethal. The birds sitting on the wire are also at a very high voltage relative to the ground. However, because they have both feet on the wire, there is no conducting path for electric current to follow to reach this lower potential – so no current flows. Meanwhile, the child flying the kite is in grave danger of providing just such a conductive path, straight through the kite and the child's body. The kite would not even have to touch the wires; as we have seen, at very high potentials current can flow even through the air if the kite comes close enough.

Like all technologies, electricity can be dangerous. The point is to use it safely, and to use it safely we need to understand it.

Electrical installations in homes tend to be fitted with a number of safety features. Figure 9.31 shows some common types of plug for mains sockets.

Notice that some of the plugs have two pins, others have three. The third pin is an 'earth' connection. This pin is connected to a wire that runs into the ground at some point – either by connecting to metal drainage piping (not ideal) or, better, to a large copper plate that is buried in the ground at a power plant. The earth connection is intended to provide an alternative conductive path for any current that has short-circuited away from the intended circuit. For example, a metal-sided toaster, or a metal lamp should have an earth connection just in case either of the other

Figure 9.28 Birds can sit unperturbed on high-voltage cables

Figure 9.29 Flying kites under high voltage cables is very dangerous indeed

two 'live' wires become loose and contact the casing – with dangerous results for anybody touching the metal casing at the time.

Another safety feature is to 'cut' the electric current the moment it becomes higher than expected. One common way to do this is to use a piece of metal that heats to melting point when a current of a certain size passes through it. This is known as a **fuse**.

ACTIVITY: Shocking danger!

In pairs, discuss the dangerous situations shown in Figure 9.30 and try to answer the following questions for each:

- **What is going on?**
- **Why do you think it is dangerous?**

Justify your reasoning with reference to what you know about current, potential difference, and resistance.

Can you think of any other similarly dangerous situations involving electricity?

■ **ATL**

- **Transfer skills**: Apply skills and knowledge in unfamiliar situations
- **Critical-thinking skills**: Evaluate and manage risk

■ **Figure 9.30** Señor Cargado's shockingly bad morning

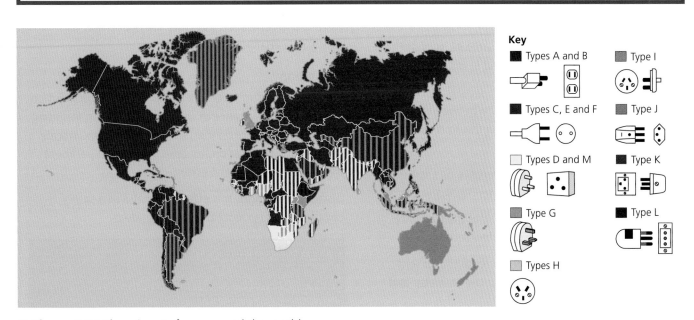

■ **Figure 9.31** Plug pinouts from around the world

Key

- Types A and B
- Types C, E and F
- Types D and M
- Type G
- Types H
- Type I
- Type J
- Type K
- Type L

ACTIVITY: Hot resistance

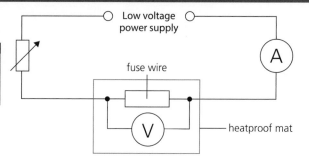

■ **Figure 9.32** Circuit diagram – fuse experiment

Inquiry: To investigate the use of fuse wire to limit electric current.

Safety: This experiment involves the use of relatively large electric currents and exposed molten wire. Use appropriate safety equipment at all times, as indicated.

Equipment

- **Fuse wire, 15 × fixed lengths**
- **Variable resistor or rheostat**
- **Power supply, 5 A maximum current**
- **Ammeter**
- **Heatproof mat or surface**
- **Tongs**
- **Wires and connecting clips**

Variables

Independent: thickness of fuse wire

Dependent: current

Controlled: length of fuse wire

Hypothesis

Suggest a relationship between the thickness of the fuse wire in a circuit and the amount of current the circuit can carry. **Justify** your hypothesis with reference to what you have learned about current and resistance.

Method

Set up the circuit as shown in Figure 9.32, so that the samples of fuse wire can be placed in series with the rheostat and ammeter.

Place one of the pieces of fuse wire between the connecting clips and then position this so that the fuse wire is over the heatproof mat.

Slowly increase the potential difference on the power supply until you see current registering on the ammeter.

Continue to increase the current until the fuse wire begins to glow, then pause for a few seconds.

If the fuse wire does not melt, increase the current by the smallest scale increment on the ammeter and pause again.

Continue until the fuse wire melts. Note the current when this occurs.

Turn off the potential difference at the power supply.

Allow the equipment to cool for two minutes, then use the tongs to remove the remains of the fuse wire.

Now repeat the process with multiple pieces of fuse wire connected in parallel. Record currents for two, three, four and five widths of fuse wire.

Results

Record your results in a suitable table and note measurement uncertainty.

Analysis

Present your results on a suitable graph. **Analyse** your graph to **suggest** a relationship between the electric current and the width of the wire. **Formulate** this relationship as an equation.

Conclusion

Explain the relationship you have found in terms of current and resistance. **Evaluate** the hypothesis you made, now that you have results.

Evaluation

Identify sources of error in the experiment. **Comment** on the relative significance of the sources of error, including measurement uncertainties, such as the minimum scale uncertainty. **Suggest** improvements to the experiment that might minimize the error.

Discuss: How might this effect be useful as a safety device to limit current? **Outline** your ideas.

◆ Assessment opportunities

This activity can be assessed using Criterion C: Processing and evaluating.

We can see how much energy is released as heat when resistance is high enough, and/or current is high enough. A **fuse** works on this basis because it melts when the current reaches a certain limiting value and so stops the current flowing. It is very important that a fuse with the right current rating is fitted to an appliance or circuit, to prevent circuit damage or injury if the current should become too high. Fuses are sometimes fitted to the devices themselves, but are usually fitted to the point where the electrical circuits of the house branch out from the electricity supply grid in a **fuse** box.

The heat is generated because electrons moving through the conductor collide with the atoms of the conducting material. When the electrons and atoms collide, the atoms gain kinetic energy from the electrons and their thermal vibration increases (see Chapter 6). In other words, the conductor begins to get hot. The more the atoms vibrate, the more likely it is that electrons will collide with them – and so the resistance of the conductor increases still further and even more heat is produced. This 'vicious circle' is an example of an **exponential relationship** and we will explore this further in Chapter 11.

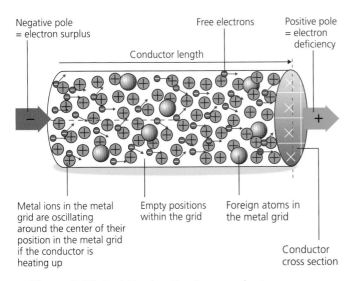

■ **Figure 9.33** Resistive heating in a conductor

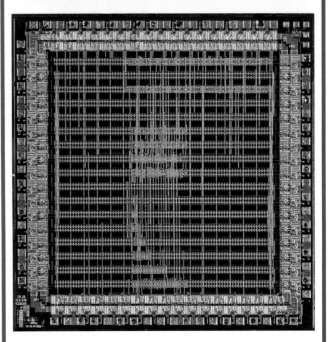

■ **Figure 9.34** Microchip interconnects

Resistive heating is a real problem in microelectronic circuitry, such as the CPU (Central Processing Unit) in your computer, tablet, or smartphone. Semiconductor devices like CPUs use electric currents that, while seemingly relatively small, are conducting through connections that are very tiny indeed.

It is extremely important, therefore, to have a clear understanding of the ways in which the physical properties of conductors affect their electrical properties.

Design an investigation using the investigation cycle to investigate the effect of one physical property of a conductor on its electrical properties.

How much does electricity cost? What is the balance between benefits and costs of electricity?

COSTING THE EARTH?

As the saying goes, nothing comes from nothing – and electricity brings not only benefits but also a cost. In Chapter 11 we will find out about the ways in which electricity is generated from natural resources, and consider environmental issues related to this. But how do we pay for the electricity?

In Chapter 6 we explored the physics concepts of work done and power, where

work done = energy changed = Pt

Since electricity also does work in a certain amount of time, we should be able to measure the amount of power used by an electrical device. This turns out to be relatively simple, since

$$I\,(A) \;=\; \frac{Q\,(C)}{t\,(s)}$$

and

$$V\,(V) \;=\; \frac{E\,(J)}{Q\,(C)}$$

then multiplying and cancelling out the charge, Q,

$$IV \;=\; \frac{E}{t} \;=\; P$$

Electrical power is measured just like mechanical power, in Watts. As we saw in Chapter 6, we can find out how much power electrical appliances use by looking at the information plaques on them.

When charging for electricity, providers usually use the amount of energy that a customer has used, which is power × time. Rather than use the Joule as a unit, electricity providers generally use the equivalent **kilowatt-hour,** or **kWh**.

ACTIVITY: Understand your electricity bill

A little physics understanding can go a long way when figuring out our electricity bills, and even further in helping us save energy, save money, and reduce our environmental impact.

Green energy

Electricity emergencies
0800 082 1234

Sheet 1 of 2

Mr B Smith
15 East Street
Plymouth
Devon
PL6 4LE

(A) Electricity bill: £123.23

Based on an **ESTIMATED** reading.

Please pay by 15 May 2015.

To pay by credit card or set up a Direct Debit call free on 0800 000 000. You can also the payment slip below.

Your bill reflects new electricity rates from 20 April 2015. The cost of each unit has risen by 7%.

(B) Your average daily electricity usage

This bill last year — 13.83 units per day
Last year — 8.8 units per day
This year — 16.43 units per day

0 5 10

Did you know?
Switching off a television, computer and other electrical equipment can reduce your electricity bill.

Bill sumamary

Last bill	£50.27
Payments	−£50.27
Charges	£123.23
Total for this period	**£123.23**

Please pay by 15 May 2015.

Please see overleaf for full details of your bill.

This is an **ESTIMATED** bill. You could save money by providing an actual reading.

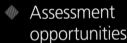

3541378974212	143 6707	£123.23

Signature _____

Date __/__/__ | 34–03–72 |

£ _____

Green energy

C6300434583116   2

C72104345501698   C3270445345000676

■ **Figure 9.35** An electricity bill

Discuss and **interpret** what each of the following points on the bill in Figure 9.35 tell us.

- Point A
- Point B

Research and find out how much a kWh of electricity currently costs, on average. Thus **estimate** how much it costs to do the following. **Justify** your estimates by **outlining** the assumptions you have made.

- **Boil a kettle of water once.**
- **Leave your computer switched on for 24 hours.**
- **Leave a light bulb switched on all night.**

Evaluate your calculations above to **suggest** three ways in which energy consumption could be reduced. **State** the order of importance of the three ways you suggested.

ACTIVITY: Power costs the Earth

Research and find one way in which our power consumption can be reduced. **Explain** how the power will be saved using physics you have learned.

Estimate how much power you could save, and how much money you could save, if your parents or carers adopted this method of reducing power consumption.

Write a report for your parents or carers. Present your findings in tables, charts, graphs that will explain to them how much energy your family is using. In your report, **evaluate** the impact on people's daily lives of making these changes and **compare** to the benefits the reduction in consumption would bring.

Here are some websites that you might find helpful:

- www.energysavingtrust.org.uk/
- http://hes.lbl.gov/
- www.aceee.org/sector/residential
- www2.ademe.fr/servlet/ getDoc?id=11433&m=3&cid=96 (in French)
- www.idae.es/index.php/lang.es (in Spanish)

◆ Assessment opportunities

This activity can be assessed using Criterion D: Reflecting on the impacts of science.

ACTIVITY: Rare metals

Electronic devices use valuable, rare metals in their construction. Research these rare earth metals and find out how they are extracted, and what happens to them when the electronic devices that contain them are discarded.

Explain the science behind the use of rare earth metals, and **discuss** and **evaluate** the implications of their use on the environment, on the global economy, and the communities affected by them.

Write a report to **summarize** your findings, including a conclusion about the impact of rare earth metals in electronics.

◆ Assessment opportunities

This activity can be assessed using Criterion D: Reflecting on the impacts of science.

SOME SUMMATIVE PROBLEMS TO TRY

Use these problems to apply and extend your learning in this chapter. The problems are designed so that you can evaluate your learning at different levels of achievement in Criterion A: Knowledge and understanding.

THIS PROBLEM CAN BE USED TO EVALUATE YOUR LEARNING IN CRITERION A TO LEVEL 1–2

1 a **Define** the resistance of a conductor.
  b **Interpret** the information in the table and **suggest** which of these wires would have the **lowest** resistance.

Wire	Material	Length (m)	Diameter (mm)
1	copper	150	0.5
2	copper	100	0.75
3	steel	100	0.75

THIS PROBLEM CAN BE USED TO EVALUATE YOUR LEARNING IN CRITERION A TO LEVEL 3–4

2 a **Outline** how the potential difference across a conductor affects the current that can flow through it.
  b **Analyse** the circuits in Figure 9.36 to **find** the missing values in each.

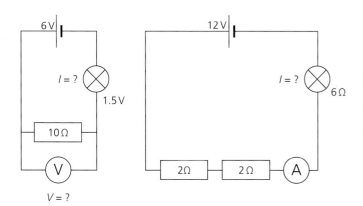

■ **Figure 9.36**

THIS PROBLEM CAN BE USED TO EVALUATE YOUR LEARNING IN CRITERION A TO LEVEL 5–6

3 Figure 9.37 shows a circuit diagram for a domestic toaster.

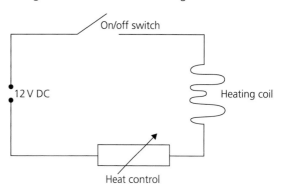

■ **Figure 9.37**

On the bottom of the toaster, there is a plate which reads 'Maximum power = 0.12 kW'.
  a **Describe** how the heat control works.
  b Using only the information given, **calculate** the maximum current the toaster draws from the mains supply.
  c The resistance of the heating coil is 500 Ω. **Calculate** the power produced as heat by the heating coil when it is set to maximum power.
  d An engineer wants to improve the design of the toaster and she thinks it will be a good idea to add a second heating coil, connected in parallel with the first. **Outline** how this innovation would change the performance of the toaster.

THIS PROBLEM CAN BE USED TO EVALUATE YOUR LEARNING IN CRITERION A TO LEVEL 7–8

4 A trans-Atlantic telephone system uses copper wire with a resistance of 0.02 Ω per metre. The distance between Kerry in Ireland and New York is about 8000 km.
  a **Calculate** the resistance of the wire used to make the telephone calls.
  b If the voltage of a signal in the wire is about 60 V in New York, **calculate** the current of the signal travelling through the wire.

c What might be **one** way to reduce the resistance of the wire? **Explain** your answer.

d **Outline** any disadvantages there might be for the telephone system in using your idea from part **c**?

The wire is now used to transmit digital information. This works by sending electrical 'pulses' down the wire from Kerry, as shown in Figure 9.38.

e On a copy of the graph, **sketch** how the resistance might affect the pulses at the end of the wire in New York. **Justify** your answer.

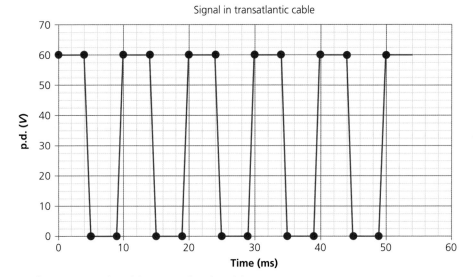

■ **Figure 9.38** Signal in transatlantic cable

Reflection

In this chapter we have found out how electricity is used to do useful work, how it is controlled, how it is measured, how we can use it safely and how much it costs. We have explored the ways in which materials conduct electricity, different kinds of circuit, and how to build safety systems into circuit design. We have also taken action to measure the cost of electricity and to find out how we can reduce our electricity consumption, our electricity bills, and so our impact on the environment.

Use this table to reflect on your own learning in this chapter.		
Questions we asked	Answers we found	Any further questions now?
Factual: How do we gain energy from electricity? How do we control electricity? How do we measure electricity? How much does electricity cost?		
Conceptual: How do electrical systems help us develop new ways of living? How can we use electricity safely?		
Debatable: What is the balance between benefits and costs of electricity? Is it realistic to think we can reduce the costs?		

Approaches to learning you used in this chapter	Description – what new skills did you learn?	How well did you master the skills?			
		Novice	Learner	Practitioner	Expert
Information literacy skills					
Communication skills					
Critical-thinking skills					
Creative-thinking skills					
Collaboration skills					
Affective skills					
Transfer skills					
Learner profile attribute(s)	Reflect on the importance of a caring attitude for our learning in this chapter.				
Caring					

⑩ Power to the people?

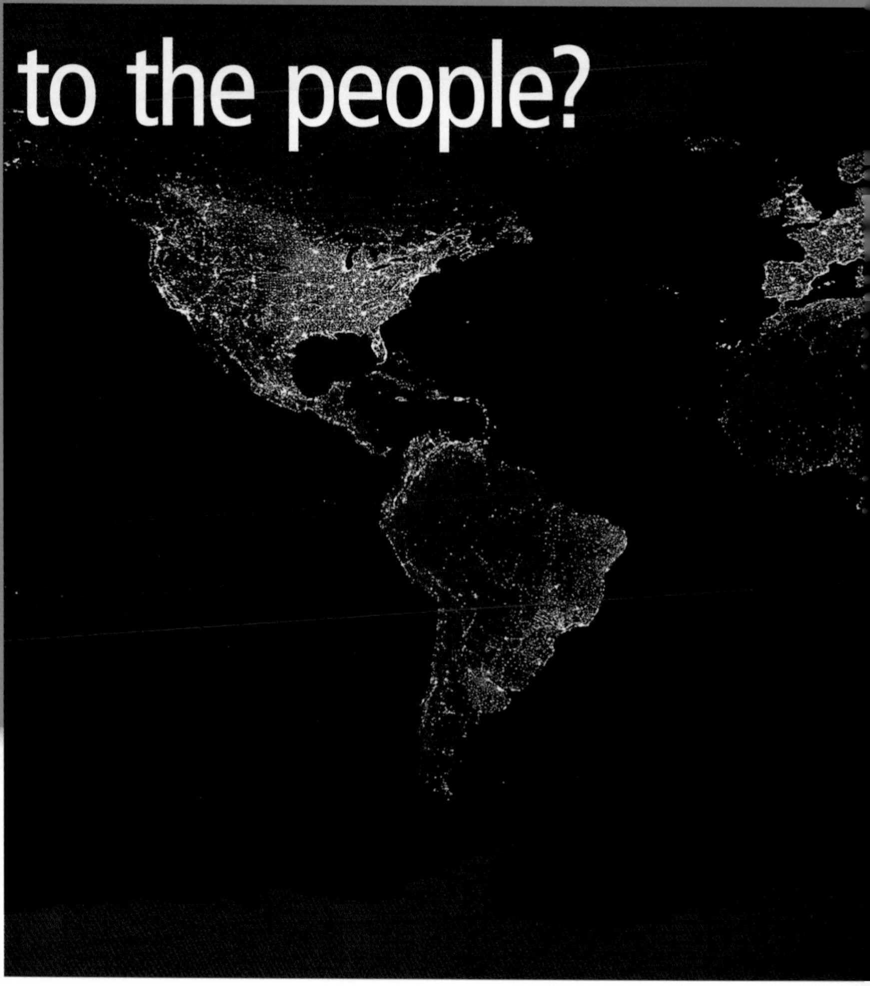

○ Manipulating the **relationship** between **interacting** electric and magnetic forces makes it possible to distribute plentiful **energy** to **everyone**.

CONSIDER AND ANSWER THESE QUESTIONS:

Factual: What are the properties of electrical fields? What are the properties of magnetic fields? What uses have we found for electromagnetic interactions? How is electricity generated? How is electricity distributed around the world?

Conceptual: How do electrical and magnetic fields interact?

Debatable: How is development around the world related to the availability of electricity? What challenges are there in achieving fair and equitable electricity distribution for everyone?

Now **share and compare** your thoughts and ideas with your partner, or with the whole class.

○ IN THIS CHAPTER, WE WILL ...

- **Find out** how electric power can be generated, distributed and controlled in different ways.
- **Explore** the advantages and disadvantages of different power distribution systems, and reflect on their impact in remote places and less economically developed countries.
- **Take action** to advocate solutions to the problem of power distribution in remote places and less economically developed countries.

■ These Approaches to Learning (ATL) skills will be useful ...

- Critical-thinking skills
- Creative-thinking skills
- Information literacy skills
- Communication skills

KEY WORDS

alternate	induce
cycle	rotate
field	transform

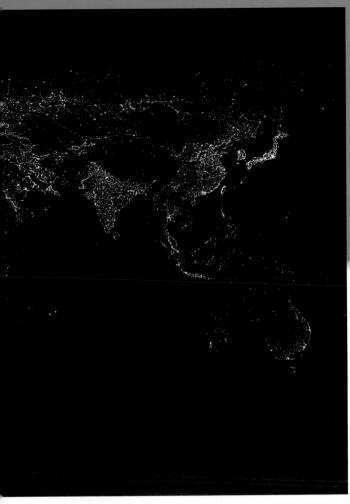

Figure 10.1 Composite map showing electrical illumination across the world at night

Figure 10.2 Quebec blackout in 1989

One night in March 1989, all the lights went out in Quebec, Canada. A huge solar storm had caused enormous changes in the Sun's electromagnetic field. The solar storm caused a surge of electric current across Quebec's power grid, which melted transformers and shut the system down for nine hours.

Have you ever experienced a power outage? If so, you will have some idea what it means to try to do even the most basic everyday things without electricity. Of course, there is no TV, no internet or telephone system – but when temperatures are freezing or lower, no electricity can also mean no heating, and during the night no lighting in homes or in the streets. Think about it this way: when is the first time you use electricity every day? The chances are, it is almost as soon as you wake up.

Electricity has become so fundamental to our lives, it takes a power outage like the one in Quebec to make us realise how much we depend on it. Electricity has become available almost everywhere in the world: but not yet everywhere.

Figure 10.1 shows electric lighting across the Earth's surface. Of course, this is a photomontage – it would be worrying if the entire Earth were experiencing night at the same time!

What do you notice about the distribution of electric light across the Earth's surface? Does everyone have the same access to electric power?

We will reflect on this learner profile attribute …

- Caring – how do people live without mains electricity? In what sustainable ways could we help them to develop it?

Assessment opportunities in this chapter

- ◆ **Criterion A**: Knowing and understanding
- ◆ **Criterion B**: Inquiring and designing
- ◆ **Criterion C**: Processing and evaluating
- ◆ **Criterion D**: Reflecting on the impacts of science

What are the properties of electrical and magnetic fields? How are they related?

INTERACTING FIELDS

ACTIVITY: Electromagnetic fields: compare and contrast

 ATL

■ **Critical-thinking skills**: Evaluate evidence and arguments

In pairs, take a sheet of paper and draw two interlocking circles to form a Venn diagram.

One person now writes down, in a space outside the circles, any knowledge about electrical fields that they can remember (check back in your work to help you recall if you need to).

The other person does the same for magnetic fields, on another part of the sheet of paper.

Now **compare** the knowledge statements.

● Write any properties or facts that are **specific to electrical fields** in one circle
● Write any properties or facts that are **specific to magnetic fields** in the other circle
● Write any properties or facts that are **shared** by the fields in the intersecting segment.

Now **evaluate** the claim that 'electrical and magnetic fields are very similar'. **Summarize** your findings in a short paragraph about electrical and magnetic fields.

In Chapter 2, we explored electrical and magnetic force fields, and discovered that – while different – they had some interesting similarities. Before we begin to explore how we can use the relationship between electrical and magnetic fields, it is a good idea to remind ourselves of those properties.

Perhaps it should not surprise us that electrical and magnetic fields are so closely related – after all, they are both caused, in some way, by the behaviour of electrons. However, the two fields are not only related – in fact, they produce one another. This almost magical connection was first identified by accident by the Danish experimenter Hans Christian Ørsted in 1820. Ørsted was giving a public lecture on the heating effect of electricity (see later in this chapter), when he suddenly thought to see if there was a magnetic field in the vicinity of his experiment. As he had a compass to hand, he placed it next to the wire – and discovered the electromagnetic interaction.

ACTIVITY: Spiralling fields

■ **Critical-thinking skills**: Practise observing carefully in order to recognize problems

Safety: As this activity involves relatively high electric currents, it may be necessary for your teacher to demonstrate it.

Equipment

- **Low-voltage, high-current power supply (at least 10 A)**
- **Copper wire of at least 0.5 mm diameter and about 1 m length**
- **Small plotting compasses × 4**
- **Lab stand and three clamps**
- **Piece of card with a 1 cm diameter hole in the middle**

Method

Set up the equipment as shown in Figure 10.3.

Safety: Note that the wire must pass through the centre of the hole in the card, so that it does not touch the card itself. Why is this? **Consider**: What happens to a wire when a lot of current passes through it?

Position the plotting compasses around the wire at right angles as shown (Figure 10.3).

Turn on the current!

Write down your observations of the effect of the current on the compasses.

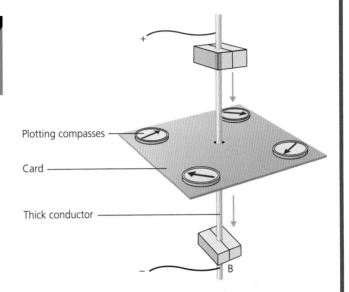

■ **Figure 10.3** Øersted apparatus

Turn off the current, then change the position of the compasses to obtain other readings for the effect.

Change the height of the card and repeat until you have an outline of the effect around the wire.

Now reverse the connections from the power supply to the wire (so that the current is flowing in the opposite direction to before). Repeat your observations.

Summarize your observations. What is happening around the current-carrying wire? What is the shape of the effect produced?

Sketch a diagram showing the effect.

ACTIVITY: Recycling fields

In Chapter 2, you may have investigated the use of electromagnets in metal-recycling plants.

We need to understand what factors affect the strength of the electromagnet, if the recycling equipment is to pick up the metals effectively.

A **solenoid** is an electromagnet arranged along a straight, ferrous bar. A solenoid can be used to pick up ferrous metals, as shown in Figure 10.5.

■ **Figure 10.5** Solenoid arranged over a conveyor

Aim: To investigate the factors affecting the strength of an electromagnet.

Use the investigation cycle to design an investigation that will enable you to measure the effect of one design factor on the strength of the magnetic field produced by a solenoid.

Plan your investigation so that you can measure the effect for various values of your chosen design factor **continuously**.

Safety: Be aware of the effects of resistive heating, as seen in Chapter 9. Design your solenoid so that the heating effect is limited. Consider how the heating effect might affect your observations, and plan your procedure accordingly.

> Hint
>
> In your evaluation, quantify the measurement uncertainties and comment on the relative importance of different sources of error. Relate your evaluation to any graphs or data you produce from the investigation.

◆ Assessment opportunities

This activity can be assessed using Criterion B: Inquiring and designing, and Criterion C: Processing and evaluating.

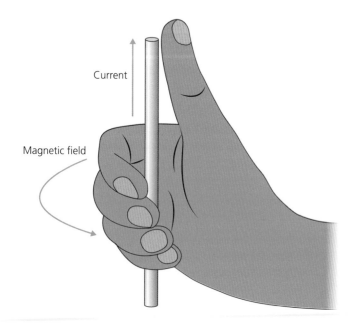

Current

Magnetic field

■ **Figure 10.4** Maxwell's corkscrew rule

The electric current produces a magnetic field that spirals around the wire. The polarity of the magnetic field depends on the direction of the electric current. One way to work out the polarity of the field is to use the 'thumbs-up' rule (see Figure 10.4), sometimes known as Maxwell's corkscrew rule, after the Scottish physicist James Clerk Maxwell.

Notice that the thumb points in the direction of the **conventional current flow** and not that of the electron motion.

Of course, as soon as the electric current is switched off again, the magnetic field disappears. This means we now have a way to produce temporary magnetic fields whose properties we can control by manipulating the current and the conductor – as opposed to a permanent field like the ferromagnetic fields we looked at in Chapter 2. **Ferromagnetic** materials will also respond to the presence of an external magnetic field and act as magnetic 'conductors', channelling and concentrating magnetic field lines through them.

What uses have we found for electromagnetic interactions?

ELECTROMAGNETS STRIKE BACK!

We have seen how moving electrons produce a magnetic field in the space around them, which can cause a magnetic force on electrons in magnetic materials nearby. This force may be strong enough to cause an object to move. If we want to make this electromagnetic interaction stronger, we can increase the strength of the electromagnetic field. Alternatively, we could introduce a second electromagnetic field, and a force will then be produced through the interaction of these two fields – just as we can double the strength of a magnetic force by adding a second magnet.

ACTIVITY: Catapulting fields

For safety reasons your teacher may need to demonstrate this effect.

Equipment

- **Large 'major' horseshoe magnet (typical field strength 0.2 Tesla between poles)**
- **Three aluminium rods, approximately 0.2 mm diameter and 20 cm long**
- **Two clamps and lab stands (preferably aluminium)**
- **Power supply, 5 A maximum current**
- **Connecting wires and clips**

Method

Set up the apparatus as shown in Figure 10.6. The aluminium bars are not attracted by the major magnet but the connecting clips and lab stand bases may be, so take care not to let them stick to the magnet!

Safety: When the current is switched on, the aluminium rods will become 'live' so take care not to touch them.

Turn the voltage on the power supply to zero, then switch the power supply on. Gradually increase the voltage and watch what happens to the aluminium bar!

Discuss what you observed. **Explain** why you think this happened.

Now use a plotting compass to identify the North and South poles of the major magnet.

Predict what will happen if we turn the magnet around, that is, reverse the magnetic poles.

Predict what will happen if we swap the connecting clips around, that is, reverse the current direction through the aluminium rods.

Test your predictions.

Suggest a use for the catapulting field effect. **Outline** limitations with this design and **suggest** what could be done to make the motion effect produced more useful.

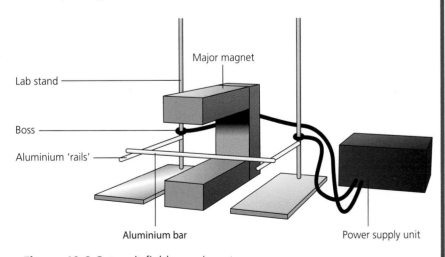

■ **Figure 10.6** Catapult field experiment

■ ATL

■ **Critical-thinking skills:** Draw reasonable conclusions and generalizations

Use the left-hand FBI rule to **explain** the motion of the rod in the catapult field demonstration. **Sketch** diagrams to show how this works for each orientation of the static magnetic field, and for each direction of current flow through the rod.

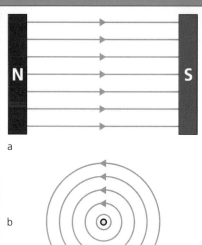

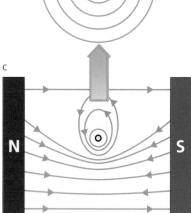

■ **Figure 10.7** Catapult field-effect diagram (a) shows the field between poles of the major magnet, (b) shows the field around the conductor, (c) shows the interaction of the two fields

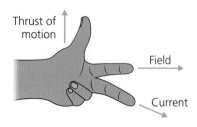

■ **Figure 10.8** Fleming's left-hand rule

The catapult effect occurs because the electromagnetic field created around the aluminium rod or wire in the circuit interacts with the magnetic field from the magnet. Of course, we could replace the permanent magnet with another electromagnet, and the effect would be the same.

The motion of the conductor is always at right angles to the current and to the magnetic field direction. In the demonstration, the permanent magnetic field (usually represented by the letter B) is arranged at right-angles to the direction of the electric current (represented as usual by the letter I). This means the conductor's magnetic field is spiralling around and interacting at right angles with the field between the magnet's poles. The consequent force between the two fields (attracting or repelling, depending on the direction of the field lines) is always then at right angles to both of these.

This complex effect can be simply visualized using a rule named after the British physicist John Ambrose Fleming (1849–1945). Fleming's left-hand rule tells us the direction of motion of the conductor (Figure 10.8).

One way to remember this is as the 'FBI' rule, where

■ F is the **force** on the conductor, represented by the thumb
■ B is the static **magnetic field** from the magnet, whose direction (North to South) is given by the index finger
■ I is the **direction of current flow** (positive to negative), given by the middle finger.

The only problem with the catapult effect is that it is rather short lived: as soon as the conductor leaves the strongest part of the magnetic field, the force drops, such that no acceleration of the conductor is produced. We have to return the conductor to its starting point for the effect to begin again. Relatively soon after Ørsted had detected the electromagnetic effect, experimenters began to try to use it to produce an 'engine' that could provide continuous motion. Michael Faraday was the first to make it really work, in 1821, with a functional prototype for a **homopolar motor**, so called because the conductor in the field always moves through the magnetic field lines in the same direction. This results in an **electromotive force** or **emf** that, in turn, produces a constant turning motion on the conductor.

While simple in design, it is relatively difficult to make homopolar motors which can produce significant amounts of **power**. A different design uses a **commutator** to produce continuous motion.

Figure 10.9 A simple homopolar motor design

To understand how the commutating motor produces continuous motion, we need to analyse again the way in which the two magnetic fields are interacting as the coil rotates. One way to do this is to use the FBI rule to visualize what would happen as a coil is rotated between two poles of a magnet, if there were no commutator arrangement.

Using the FBI rule, we can see that in the 'flat' or zero degree position (Figure 10.10), the current is travelling into the page on the left-hand side of the coil, and out of the page on the right-hand side of the coil. This produces force in opposite directions, causing the coil to rotate in an anti-clockwise direction.

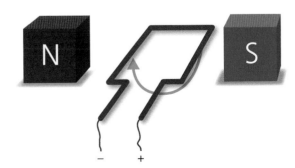

Figure 10.10 End view of spinning coil, no commutator, 0°, + left, N left

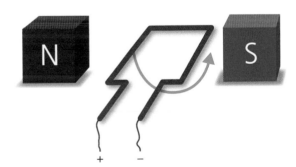

Figure 10.11 Spinning coil, no commutator, 180°, + right, N left

When the coil has flipped over through 180°, however, the wires at the end of the coil have crossed (Figure 10.11). This means the current is now flowing into the page on the right-hand side of the coil and out of the page on the left-hand side of the coil. We still have opposing forces, but they are now pointing in the opposite direction to before. The coil would just flip back on itself – we would have a rocking motion, rather than a continuous rotation.

The commutator effectively breaks the circuit every half-rotation, so that the current is always travelling into the page on the left-hand side and out of the page on the right-hand side. The opposing forces on each side of the coil are always causing the coil to move in the same anti-clockwise direction (Figure 10.12).

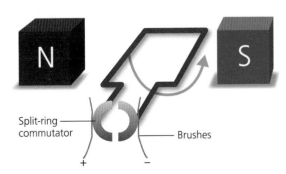

Split-ring commutator — Brushes

Figure 10.12 Spinning coil, commutator, 0°, + left, end view

ACTIVITY: Electromagnets in motion – making stronger motors

■ **ATL**

■ **Creative-thinking skills:** Apply existing knowledge to generate new ideas, products or processes

In pairs, brainstorm some of the limitations of the two motor designs you looked at in the 'Activity: Electromagnets in motion – comparing motor effects'. **Record** your ideas in the first two columns of a table like this:

Design feature	Limitation it produces	Improvement

In a third column **suggest** how the motor designs might be improved.

Figure 10.13 shows a different commutating motor design.

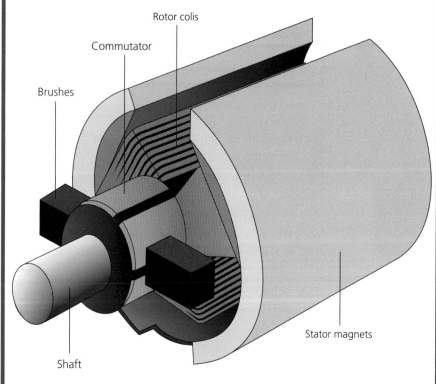

■ **Figure 10.13** Commutating dc motor, multiple coils and commutators

Interpret the diagram to **identify** how the design of this motor is different to the simple commutating dc motor (page 193). **Suggest** how each design modification might improve the motor, and **explain** using the FBI rule for motor effect.

◆ Assessment opportunities

In this activity you have practised skills that are assessed using Criterion A: Knowing and understanding.

ACTIVITY: Electric motor energy

■ **ATL**

■ **Creative-thinking skills:** Apply existing knowledge to generate new ideas, products or processes

In Chapter 6, we **analysed** work done by machines in terms of the energy changes they cause. A **Sankey diagram** was one way to do this.

Complete the Sankey diagram (Figure 10.14) for an electric motor.

Now redraw the diagram **backwards**, starting with the output and then reversing the direction of the arrows.

Outline what might this machine do?

MAKING A SPARK

In Chapter 9 we saw that large electrical charges can be generated using friction between conductors. Unfortunately, the energy held in machines like the Van der Graff generator is quite small and we can't use the static electricity stored to useful effect. So how is useful electricity made?

HOW IS ELECTRICITY GENERATED?

If – as Ørsted showed – an electric current could be used to produce a magnetic field, could a magnetic field be used to produce an electric current? The first person to test this **hypothesis** was Michael Faraday in 1831.

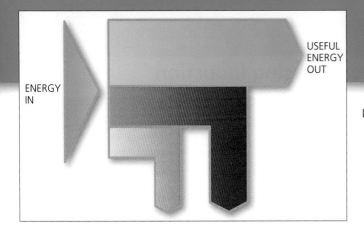

ENERGY
IN

USEFUL
ENERGY
OUT

■ **Figure 10.14** Sankey diagram

ACTIVITY: Making electricity

Write a hypothesis about the creation of electric current using electromagnetic fields, referring to what you have learned about electromagnetic effects.

Observe what happens in the following experiment.

Equipment

- **A strong bar magnet**
- **A solenoid coil**
- **A sensitive ammeter (or 'galvanometer') that can detect current of approximately 5 mA**
- **A spring**
- **Lab stand and clamp**
- **Small masses and mass hanger**

Set up the apparatus as shown in Figure 10.15.

Add masses to the mass hanger, so that the magnet is at equilibrium in the centre of the coil.

Pull the masses down and release the system.

Observe what happens to the galvanometer needle.

Experiment with moving the magnet at different speeds and **observe** the size and the direction of the effect on the galvanometer needle.

Conclusion

Outline your observations. To what extent was your hypothesis correct?

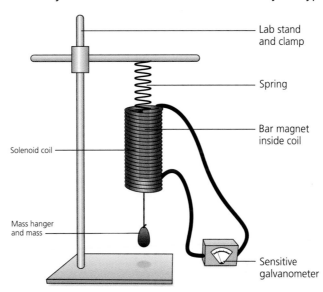

Lab stand and clamp

Spring

Bar magnet inside coil

Solenoid coil

Mass hanger and mass

Sensitive galvanometer

■ **Figure 10.15** Electric current can be produced by moving a magnetic field through a coiled wire

---

EXTENSION

Explore further! If you have an **oscilloscope** available, connect it to the coil instead of the galvanometer. **Observe** the effect on the oscilloscope trace as the magnet moves up and down through the field.

Use a **datalogger** to record the current and then produce a graph of the data.

ACTIVITY: Analysing induction

Use the right-hand FBI rule to **explain** the direction of the electric current produced. Develop your **conclusion** from the moving magnet experiment above. Thus, **explain** what happens when the magnet is stationary inside the solenoid coil.

As magnetic field lines move through the wire in the solenoid, they cause electrons to move and create an electric current. This phenomenon is called **electromagnetic induction**. Just as with the motor effect, three variables interact – motion, magnetic field, and the electric current produced. The FBI rule for motor effect works for electromagnetic induction too, except that, since we have reversed the process, we now have to reverse the geometry of the situation and use the **right** hand instead.

Again, we can make the induction effect continuous by using an electric motor design.

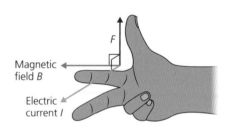

■ **Figure 10.16** FBI rule for induction – right-hand rule

ACTIVITY: Making continuous current

Aim: To observe the electric current produced by an electric motor used as a generator.

Equipment
- **Electric motor (dc)**
- **Lab stand and clamp**
- **Pulley**
- **Small masses and hanger**
- **Light bulb, 6–9 V**
- **Datalogger and current sensor**

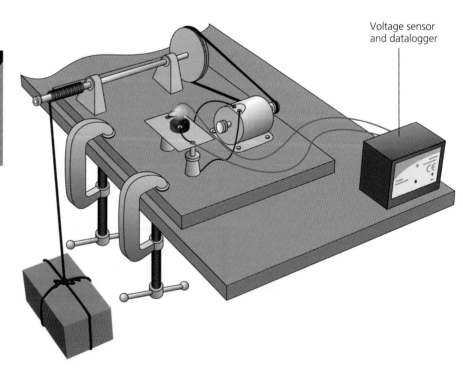

■ **Figure 10.17** Using a falling mass to drive a dc dynamo and generate electric current

The current produced in an electric motor used as a generator flows in one direction only and so is known as **direct current** or **dc.** Without a commutator, the electric current changes direction every half-turn of the coil in the generator. This produces an electric current which looks like a waveform, as we saw in Chapter 7.

Since the electric current changes direction it is called an **alternating current** or **ac** for short. The generators used in power stations work this way, and so the electricity we receive through the grid system is ac. Instead of a **commutator**, these generators allow the current to flow continuously from the coil using **slip rings** (Figure 10.18).

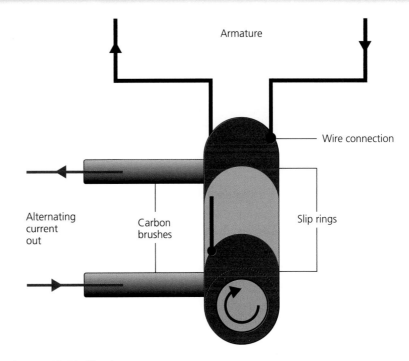

■ **Figure 10.18** Slip rings on an ac generator

Set up the apparatus as shown in Figure 10.17.

Make sure that the masses can fall some distance – perhaps the height of a lab bench.

Release the masses and log the output from the motor. Use datalogging software or a spreadsheet program to produce a graph of the output.

Analysis

Describe what happens to

- **the size**
- **the direction**

of the electric current as the motor turns.

Conclusion

Use the right-hand FBI rule to **explain** the direction of the current produced.

Compare the output from the dc dynamo to the output from the moving magnet experiment (page 195).

Remember that the electric motor has a **commutator** to ensure that the coil always moves in the same direction. **Explain** how this has affected the output from the motor. **Suggest** what design modification would need to be made in order to allow current to flow in both directions through the external circuit.

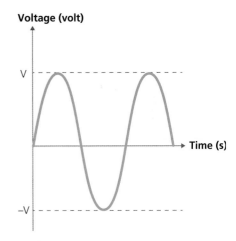

■ **Figure 10.19** ac output

How is electricity distributed around the world?

Figure 10.20 High-tension power lines in Australia

POWER TO THE PEOPLE

Electric power lines are part of the landscape in many places in the developed world. In some places the 'pylons' used to carry them seem to stride over the landscape like giants, while in others power lines may just be a single bundle of cables strung between poles. We saw in Chapter 8 how energy resources are harnessed and converted into electrical energy – the next challenge is to distribute this energy to wherever it is needed. The system for doing this is often called an **electrical grid**.

Figure 10.21 Power line spaghetti

ACTIVITY: Thinking about transmission – part 1

ATL

■ **Creative-thinking skills**: Practise visible thinking strategies and techniques

Look carefully at the information in the diagram of a power transmission grid, and **compare** to the images of power cables on the page opposite.

● **What do you see?**
● **What do you think about that?**
● **What does it make you wonder?**

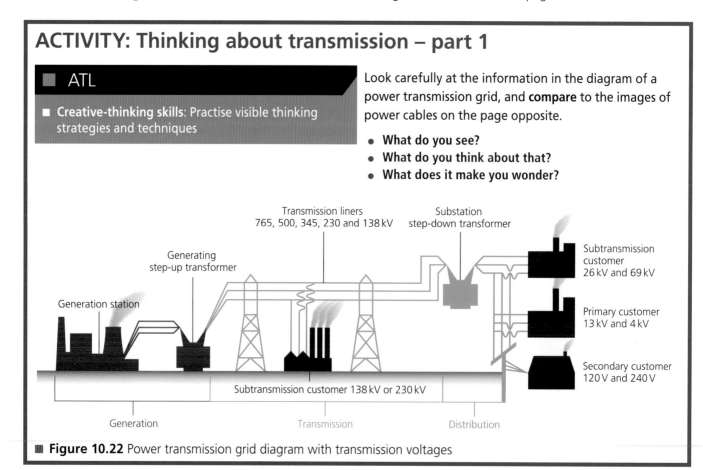

■ **Figure 10.22** Power transmission grid diagram with transmission voltages

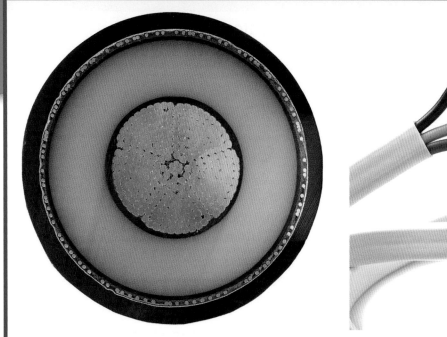

■ **Figure 10.23** Section of a high-tension cable

■ **Figure 10.24** Cable in a household mains plug

ACTIVITY: Thinking about transmission – part 2

Work in pairs.

On a sheet of paper draw a Venn diagram with two interlocking circles.

Think about the problems of using high electric currents and of using high electric voltages.

- **In one circle, one person writes all the disadvantages of using large electric currents.**
- **In the other circle, one person writes all the disadvantages of using large voltages.**

Now **share** your ideas. Are there any disadvantages which are common to both? If so, write these in the interlocking segment.

Discuss: If a national government sought your advice on the construction of a new electrical grid, what would you tell them? Is it better to use high current, or high voltage? **Summarize** your ideas in a paragraph.

In Chapter 9, we also saw how power for an electric system can be calculated from

$$P = IV$$

The energy delivered to workplaces and homes is measured in **kilowatt-hours**. The greater the amount of energy required, the greater the power that must be available in the transmission grid. In turn, we can see that this means that the product IV must increase – so more current must flow, at a higher electrical potential.

We also saw in Chapter 9 that large currents have an unfortunate side-effect in even the best metal conductors: they produce **resistive heating**. This means that electrical energy transmitted with high currents will tend to produce a lot of wasted heat, and the system rapidly becomes very inefficient, especially if we have to transmit over many hundreds of kilometres. Finally, big currents need thick wires – and that means greater economic cost. For these reasons, electric power is transmitted – wherever possible – at a very high voltage, sometimes called **high tension**. By maximizing the voltage, the electric current flowing in the wire for a given power can be minimized.

As you can see from Figure 10.22, transmission voltages can be in the kilovolts – which could present some safety issues if these the current was delivered to homes at these voltages! The grid has to incorporate a system for **stepping up** and **stepping down** the voltage to the optimal value, and to do this **transformers** are used.

ACTIVITY: Transforming electricity – invisible motion

Aim: To change the potential difference of an electric current using electromagnetic induction.

A **transformer** is a device that is used to change the voltage of an electric current. In this experiment we will investigate how a transformer works. What affects the voltage over the output of a transformer?

Equipment

- **C-shaped steel cores**
- **2 × 1 m insulated steel or copper wire**
- **Sticky tape**
- **ac power supply, set to 3 V**
- **Battery pack, 3 V**
- **2 × light bulbs, 6 V**
- **Sensitive voltmeter, data logger with voltage sensor, or oscilloscope**

Experiment 1

Wrap the wire around the C-core arms as shown in Figure 10.25. Count the turns of wire on each core and make sure they are the same. Use the sticky tape to hold the coils in position.

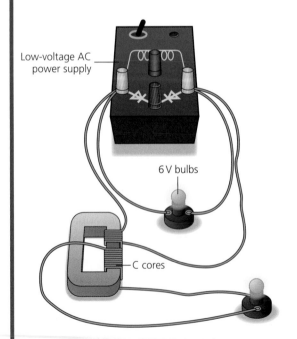

Low-voltage AC power supply

6 V bulbs

C cores

■ **Figure 10.25** Transformer experiment

Connect the apparatus as shown in Figure 10.25.

Safety: Ask your teacher to check your circuit before you turn on the power.

Place the C-cores so that their arms contact each other. Turn on the power. **Observe** the output from the second coil.

Explain how the current is produced at the output of the second coil, when there is no electrical connection between the coils. **Consider:** What is produced around the first coil? What will this produce in the second coil?

Now replace the ac power supply with the 3 V battery pack. **Observe** the output on the second coil.

Experiment 2

Independent variable: distance between the cores.

Before you begin, write a **hypothesis** about the effect of changing the distance between the cores.

Experiment 3

Independent variable: number of coils on the secondary core.

Before you begin, write a **hypothesis**. Present your results and **analyse** them.

Conclusion

Outline the effect of each of the variables. **Interpret** your results to identify any trends or patterns. **Evaluate** the validity of your hypotheses for experiments 2 and 3 – to what extent were you correct? **Explain** the difference in the outputs obtained with reference to what you know about **electromagnetic induction**.

Evaluation

State the measurement uncertainties in your results. **Comment** on the significance of the uncertainties in your experiment. **Suggest** how to make the experiment more accurate. **Outline** ways in which the design of the transformer might waste energy.

◆ Assessment opportunities

This activity can be assessed using Criterion C: Processing and evaluating.

In this activity you have practised skills that are assessed using Criterion B: Inquiring and designing.

The output voltage from a transformer depends on the ratio of the coils on the primary and secondary cores. Transformers can be very efficient (see Chapter 6), but they do waste energy, especially as heat and sound. One major design challenge in producing an efficient transformer is to reduce the **eddy currents** that tend to be produced by the alternating electromagnetic field in the cores themselves.

ACTIVITY: Going further – eddy currents, contactless braking, and super-efficient vehicles

■ ATL

- **Information literacy skills**: Access information to be informed and to inform others

Research using eddy currents electromagnetic braking to find out how electromagnetic induction can be used to slow moving objects.

Some 'hybrid' vehicles use this system to recycle kinetic energy when the vehicle brakes. **Summarize** with reference to what you have learned about **electromagnetic induction** how this technology works.

ACTIVITY: Skipping the grid

▌ Take action

! **Find out about local electricity generation.**

■ ATL

- **Information literacy skills**: Access information to be informed and to inform others
- **Communication skills**: Use appropriate forms of writing for different purposes and audiences

Research the alternative of 'skipping the grid' or using local power to provide electricity to remote places.

Explain how science you have studied is used to find a solution to the problem of power distribution, and **compare** the use of a grid system to the use of local power stations. **Evaluate** each alternative in terms of its advantages and disadvantages.

Identify one region of the world that does not currently have electric power. Apply what you have learned to **outline** how each of the solutions – local or grid power – might be used in this place.

Write a report for the local government in that region, **explaining** your research and advocating the solution you think will be most effective in bringing power to the people who live there. Then write an email which will be sent with the report, but which **summarizes** your main findings in an attention-catching way!

Be sure to **document** all sources accurately.

◆ Assessment opportunities

This activity can be assessed using Criterion D: Reflecting on the impacts of science.

SOME SUMMATIVE PROBLEMS TO TRY

Use these problems to apply and extend your learning in this chapter. The problems are designed so that you can evaluate your learning at different levels of achievement in Criterion A: Knowledge and understanding.

THIS PROBLEM CAN BE USED TO EVALUATE YOUR LEARNING IN CRITERION A TO LEVEL 3–4

1   In a recycling factory, an **electromagnet** is used to separate different metals found in metal waste.

The list in the box shows the materials found in the waste.

iron	aluminium	copper	steel	tin

a   **State** which materials will be attracted to the electromagnet.

b   **Explain** your choice(s).

c   **State** which of the diagrams A–D shows the **correct** shape of a magnetic field made around a solenoid.

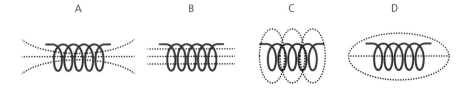

■ **Figure 10.26** Solenoid fields options A–D

An engineer wants to increase the strength of the electromagnet.

d   **State** which of the following modifications to the design would increase the strength.

   A   Use ac instead of dc through the electromagnet

   B   Use a longer, thicker core

   C   Join the two poles of the electromagnet with an iron bar placed across them

   D   Increase the number of turns on the electromagnet

e   **Outline** why one of the modifications you rejected would not increase the strength of the electromagnet.

2   The diagrams below show the orientation of the coil of a dc electric motor between two bar magnets.

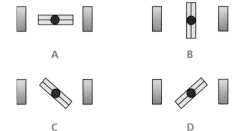

■ **Figure 10.27** dc motor coil options A–D

a   **State** which position gives the maximum force on the coil.

b   **Explain** your reasoning.

c   **Outline** how the force on the coil changes as the coil rotates between the magnets.

3   The graph in Figure 10.28 shows the variation in potential difference (p.d.) of
    an alternating current (ac) produced by a rotating generator.

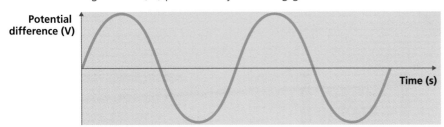

■ **Figure 10.28** ac p.d. cycle

The ac completes one complete **cycle** every 5 seconds.

a   Alex puts this ac through a centre-zero dc voltmeter. **Describe** the motion
    of the needle for one complete cycle of the ac.

b   Alex now connects a light bulb to the ac supply. **Describe** what he will
    see for one cycle.

c   Alex now causes the generator to turn 10 times faster than before.
    Outline or **sketch** the new ac output.

d   **Describe** what Alex will now observe while watching the bulb connected
    to the ac supply. **Explain** your answer.

e   With reference to the cycle above, **state** whether the ac generator Alex
    is using has a commutator. **Explain** your answer, and **suggest** what
    alternative arrangement the ac generator may have to carry current from
    the coil.

4   A length of flexible, slack copper wire is fixed so that part of it is held
    vertically in the field of a horseshoe magnet (Figure 10.29).

a   i   **Sketch** what the wire might look like when a large current passes
            through it.

    ii **Explain** why the wire looks like this.

b   i   On the same diagram, **draw** what the wire might look like if the
            current in **a** is reversed.

    ii **Explain** why the wire looks like this.

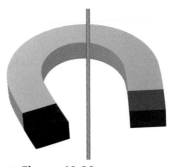

■ **Figure 10.29**

5 The diagram in Figure 10.30 shows one design for a demonstration electric motor.

   a **Label** the diagram with the parts of the motor listed in the box below.

   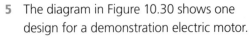

magnets	coil	axle
brushes	commutator	

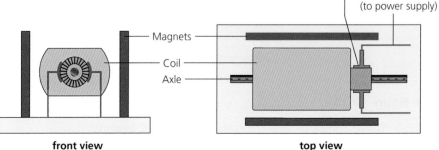

   front view   top view

   ■ **Figure 10.30**

   b **Suggest two** ways in which the amount of force produced by the motor could be increased. **Explain** how each of these modifications will increase the force produced by the motor.

   Jacopo notices that sometimes the motor stops and the coil gets 'stuck' in one position.

   c Which position is this most likely to be? **Describe** with words or a diagram to show this.

   d **Explain** your answer to **c**.

   e **Suggest** a modification Jacopo could make to the motor, so that it is less likely to get 'stuck' and provide more force (you may use a diagram if you prefer).

THIS PROBLEM CAN BE USED TO EVALUATE YOUR LEARNING IN CRITERION A TO LEVEL 7–8

Power generation using wind turbines

6 The diagram in Figure 10.31 shows the output from a wind turbine on a windy day.

   a If the wind speed increases, the turbine turns faster. **Sketch** a new curve to show how the output might change.

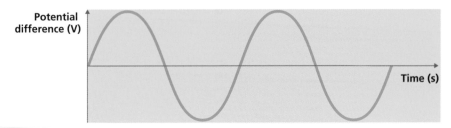

■ **Figure 10.31**

The charts that follow show the output for two different types of turbine: a 'lateral' type, as shown below, and a 'vertical' type, shown on the next page.

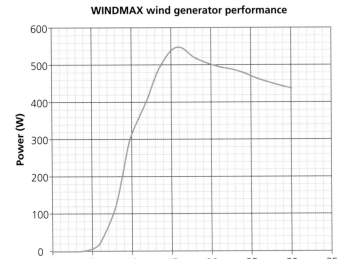

WINDMAX wind generator performance

■ **Figure 10.32** Lateral turbine performance curve

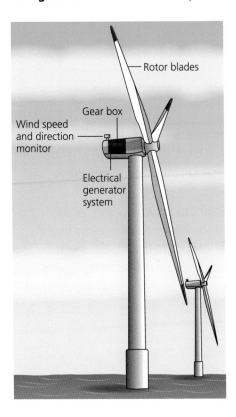

- Rotor blades
Gear box
Wind speed and direction monitor
Electrical generator system

■ **Figure 10.33** Lateral turbine

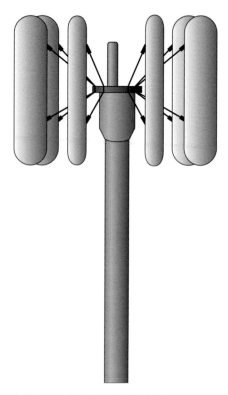

Figure 10.34 Vertical turbine

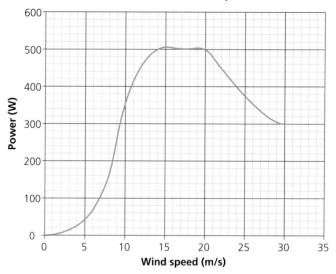

Figure 10.35 Vertical turbine performance curve

b For each of the turbines,
  i **state** the wind speed at which each turbine works best. (This is called the peak performance.)
  ii **describe** what happens to the power produced after the peak performance.

An engineer suggests three methods for increasing the power produced by the lateral wind turbine:

- Method 1: Make the turbine blades (propellers) longer
- Method 2: Increase the number of turns on the coils in the generators
- Method 3: Place the turbine somewhere where the wind blows more consistently.

c **Sketch** three graphs to show how each of the improvements the engineer has suggested might affect the performance of the wind turbine. Label each graph clearly with the method it represents.

d **Explain** the shape of the graphs you have sketched.

Reflection

In this chapter, we have explored the way in which electromagnetic fields interact to produce electrical energy. We have reflected on the relationship between electrical power distribution and development in different parts of the world, and we have considered the strengths and limitations of power grids.

We have seen in this chapter how electrical power can be distributed from central power stations. But in many parts of the developing world, the challenge of connecting remote places to the grid is considerable. One alternative to this is to generate more power locally using small, cost-effective generators of different kinds.

Use this table to reflect on your own learning in this chapter.					
Questions we asked	Answers we found		Any further questions now?		
Factual: What are the properties of electrical fields? What are the properties of magnetic fields? What uses have we found for electromagnetic interactions? How is electricity generated? How is electricity distributed around the world?					
Conceptual: How do electrical and magnetic fields interact?					
Debatable: How is development around the world related to the availability of electricity? What challenges are there in achieving fair and equitable electricity distribution for everyone?					
Approaches to learning you used in this chapter	Description – what new skills did you learn?	How well did you master the skills?			
		Novice	Learner	Practitioner	Expert
Critical-thinking skills					
Creative-thinking skills					
Information literacy skills					
Communication skills					
Learner profile attribute(s)	Reflect on the importance of being caring for our learning in this chapter.				
Caring					

11 What's in an atom?

○ Learning to control nuclear **changes** allows us to use matter *in new ways* and release huge quantities of **energy**, with **consequences** that can be both positive and negative.

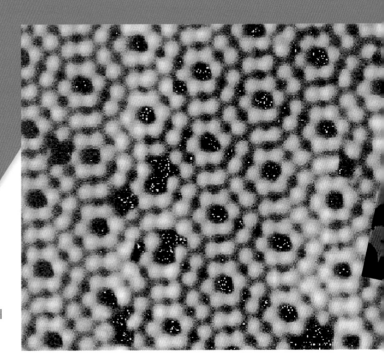

■ **Figure 11.1** Atoms in a new high-performance metal alloy, photographed with a scanning tunnelling electron microscope (STEM)

CONSIDER AND ANSWER THESE QUESTIONS:

Factual: What's inside an atom? What happens when a nucleus changes? What effects does radiation have? How can we harness energy inside an atom?

Conceptual: What new possibilities does nuclear technology offer? What special consequences does nuclear technology bring?

Debatable: Is nuclear power the answer to our sustainable energy needs?

Now **share and compare** your thoughts and ideas with your partner, or with the whole class.

○ IN THIS CHAPTER, WE WILL …

■ **Find out** the physics of the atom's interior, how the nucleus is structured and what its component parts are. We will characterize the properties of different ionizing radiations, and analyse what happens in the nucleus when they are released.

■ **Explore** the technologies humanity has devised to control and utilize nuclear reactions to produce energy.

■ **Take action** to decide whether nuclear power is the solution to our sustainable energy needs.

KEY WORDS

atom	energy	nuclei
decay	nucleus	reaction

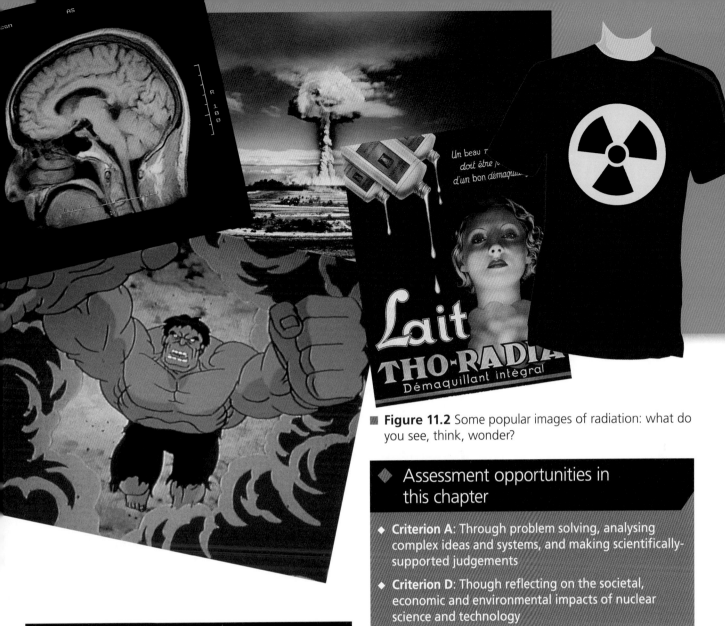

Figure 11.2 Some popular images of radiation: what do you see, think, wonder?

◆ Assessment opportunities in this chapter

◆ **Criterion A**: Through problem solving, analysing complex ideas and systems, and making scientifically-supported judgements

◆ **Criterion D**: Though reflecting on the societal, economic and environmental impacts of nuclear science and technology

SEE-THINK-WONDER

Look at the images in Figure 11.2. **In pairs, discuss:**
● **What do you see?**
● **What do you think about them?**
● **What do they make you wonder?**

We started out in Chapter 1 by thinking about the scale of things in the Universe, and we explored some different ideas about the nature of matter. In this chapter, our focus will be on the nuclear scale – to understand the physics of the interior of the atom. All of the images above relate to nuclear physics, and we can see that perceptions of this part of physics can vary widely. Like all changes, nuclear physics and its technology have brought new possibilities and also new consequences that we need to balance.

■ These Approaches to Learning (ATL) skills will be useful …

■ **Critical-thinking skills**

■ **Information literacy skills**

■ **Media literacy skills**

■ **Affective skills**

■ **Communication skills**

■ **Collaboration skills**

● We will reflect on this learner profile attribute …

● Being open-minded – we seek and evaluate a range of points of view.

What's inside an atom?

INSIDE THE ATOM

While Parmenides and Democritus might have disagreed about the 'fundamental' nature of matter (see Chapter 1), at the scale of nuclear technology, matter is understood to be discontinuous, consisting of tiny particles called **atoms**. When Lavoisier demonstrated that all materials could be reduced to 'pure' elements, it was Democritus' idea of a unitary, indivisibly solid piece of 'stuff' that was first adopted as the model of the atom. Few people had any idea just how small these things were – although it was possible to make a first approximation to get an idea.

The 'oil patch' method was first conceived as a way to measure molecule size by the British physicist Lord Rayleigh (1842–1919) in the 19th century CE, although the American Benjamin Franklin (1706–1790) made a similar observation in 1757 when observing oil on the surface of a pond in Clapham, when he was visiting London.

ACTIVITY: Measuring atoms

■ ATL

- **Information literacy skills**: Collect, record and verify data
- **Critical-thinking skills**: Practise observing carefully in order to recognize problems
- **Critical-thinking skills**: Use models and simulations to explore complex systems and issues

In this activity we will use a simple method to make an approximate measurement of the size of a molecule of oil, and from that we will try to deduce the size of an individual atom within the oil.

The experiment requires careful practical manipulation skills and close observation.

Safety: The experiment requires the use of molten wax to coat the tray – this is best done beforehand by your teacher or a lab technician.

The pollen powder can trigger allergies or asthma – if susceptible to these conditions, your teacher should check the health and safety guidance where you live.

Background

Olive oil molecules consist of 12 carbon atoms in a straight chain. One end of the molecular chain is 'hydrophobic' (or water-repelling) and the other 'hydrophilic'. This means that the water-repelling end will orientate vertically up, away from water, and when the oil forms a puddle on the surface of water the molecules will stand on end. This also means that the oil will spread across the surface of the water until it forms a layer that is only one molecule thick.

Equipment

- **Large (minimum 50 cm × 30 cm × 5 cm) tray**
- **Spirit level**
- **Wooden or plastic sticks × 2, as long as the tray is wide**
- **Paraffin wax and paintbrush**
- **Olive oil with drop dispenser**
- **Fine pollen powder (lycopodium)**
- **Millimetre-scale ruler**
- **Vernier callipers**
- **Loops of thin metal wire**
- **Magnifying eyepiece**

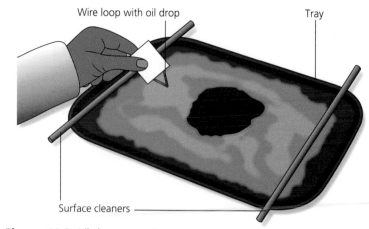

Wire loop with oil drop

Tray

Surface cleaners

■ **Figure 11.3** Oil drop experiment

Method

Before beginning, your teacher or lab technician will have prepared the sides of the tray by coating them with melted paraffin wax. This helps reduce surface tension effects at the edges of the tray.

Place the tray on a bench, preferably with one end over a sink. Use the spirit level to make sure the tray is perfectly level in both directions. Use small pieces of card as wedges to adjust the level of the tray if necessary.

Now carefully pour water into the tray (clean tap water is usually fine). Fill the tray to the very top.

Use the sticks to 'sweep' across the surface of the water, pushing excess water over the edges of the tray and cleaning the surface of dust.

Gently sprinkle the pollen powder over the surface of the water, so that it forms a very fine but visible layer.

Take the wire loop and dropper, and place the smallest amount of olive oil in the loop that you can. Use the magnifying eyepiece to measure this droplet against the millimetre scale. Use a lab stand to hold the wire and position the scale while you do this. Use absorbent paper towels or lab filter paper to remove excess oil until you have a droplet that is approximately 0.5 mm in diameter. This is diameter d.

Lightly touch the oil drop to the water in the centre of the tray.

Results

Observe the patch formed by the oil on the surface of the water. After some time, the patch should stop growing and its (approximately circular) diameter will stabilize.

Measure the diameter (D) of the oil patch with the millimetre ruler or the Vernier callipers. If the patch is not properly circular, measure major and minor diameters.

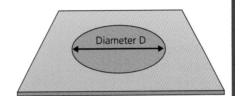

■ **Figure 11.4** Oil patch geometry

Analysis

With reference to the diagram and your measurements:
- **estimate** the volume of oil in the drop from the diameter d
- **estimate** the depth h of the oil patch from the diameter D.

> **Hint**
>
> The volume of a sphere is
> $$V = \frac{4}{3}\pi r^3$$
> The volume of a cylinder is
> $$V = \pi r^2 h$$

Conclusion

Estimate the size of a single atom in the olive oil molecule chain.

Evaluation

Research the range of accepted sizes for atoms. **Compare** the 'literature' (accepted) values with your own estimation. How accurate was your estimate?

First-order approximations from observations

In this activity we have made an approximate measurement. While not accurate, the estimate by Rayleigh of the size of molecules – and later atoms – at least gave a 'ball-park' figure that scientists could work with. Making rough initial estimations of this type is an important skill in physics, since it provides the basis for more precise experiments and measurements later on.

though the drawer was dark. Becquerel had discovered radioactivity and was later awarded the Nobel Prize for suggesting that X-rays were something different to the 'rays' that were released from fluorescent materials.

The French experimental scientists Pierre (1859–1806) and Marie (1867–1934) Curie first began to measure the properties of these mystery rays, and noticed that sometimes they produced **ionization**. Marie Curie was awarded the Nobel Prize in 1911 for her discovery of two elements which she termed **radioactive**: radium and polonium (after her native Poland). Tragically, Marie Curie also inadvertently discovered another property of radioactive materials – she died at quite a young age of aplastic anaemia, a condition often caused by exposure to large quantities of radiation.

Meanwhile, other scientists – both physicists and chemists – were exploring the properties of elements. In 1895, the German physicist Wilhelm Roentgen (1845–1923) noticed that some materials could be made to produce 'fluorescence' – a phenomenon whereby a substance produces light. At first, he concluded the fluorescence was caused by energy he mysteriously called 'X-rays'.

The French physicist Henri Becquerel (1852–1908) wanted to investigate these mysterious rays using plates of glass coated with photographic chemicals. He intended to expose some fluorescent salts to sunlight and see what happened using his photographic plates, but on the appointed day for his experiment the weather in Paris was rather dull. So, for the time being, he stuck the plates and the salts together in a desk drawer. A few days later, on taking them out of the drawer, Becquerel was astonished to see that the plates all held a 'shadow' or image of the salts on them – even

Around the same time, the New Zealander Ernest Rutherford (1871–1937) was experimenting at the University of Manchester, England, to see if the 'rays' could be used to produce a more accurate measurement of the size of atoms. He supervised his assistants Geiger and Marsden in building an experiment in which a certain kind of positively charged 'alpha ray' was fired at a thin sheet of gold leaf. The idea was that the atoms would deflect the rays and, by measuring the angles at which they were deflected, it would be possible to deduce the size of the gold atoms.

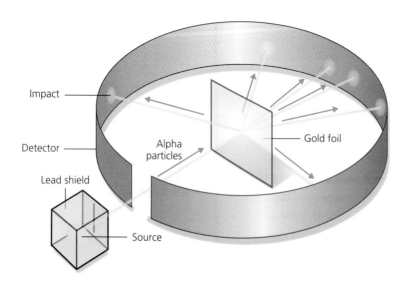

■ **Figure 11.5** Representation of the alpha-scattering experiment with which Rutherford discovered the nucleus

ACCIDENT AND DISCOVERY

■ ATL

- **Affective skills:** Demonstrate persistence and perseverance; Practise 'bouncing back' after adversity, mistakes and failures; Practise dealing with disappointment and unmet expectations

On reading through the story of the discovery of radioactivity and the nucleus, how many of the discoveries were actually the result of unexpected results and 'accidents'?

In science, accident is sometimes as important as planning. But these discoveries were made because the scientists didn't despair and give up, nor did they ignore or eliminate unexpected results because they did not fit their hypotheses. Instead, they responded to the unexpected with curiosity, and by asking why it occurred.

The alpha rays would be detected as flashes of fluorescence in a coating on the outside of the experiment vessel, and Rutherford's hypothesis was that the flashes would be seen in the areas to either side and behind the gold leaf – rather as a football rebounds from a goal post into the crowd behind the goal. But Geiger and Marsden, while calibrating the equipment one day, made an extraordinary discovery: they observed alpha-ray 'events' back near the source itself, at unexpectedly huge angles of deflection. Rutherford remarked of this discovery, 'It was as if you fired a fifteen inch naval shell at a piece of tissue paper and the shell came right back and hit you.'

When Rutherford did the analysis on Geiger and Marsden's data, he realized that it was physically impossible to explain the large-angle deflections of a few alpha particles using the conventional model of the atom. At the time, atoms were still held to be a uniform sphere of matter made from a 'mixture' of evenly distributed positively charged material with negatively charged electrons scattered through it – rather like a 'plum pudding' or a blueberry muffin.

Rutherford reasoned that the only way that such a large deflection could occur was if the atom actually contained a very large positive charge to repel that of the alpha particles.

■ **Figure 11.6** A blueberry muffin – model for an atom?

On the other hand, most of the alpha particles **did** behave as expected and passed through to the other side of the gold leaf. Rutherford had to combine the two observations in a new explanatory model. He concluded that, in fact, the atom was not an electrically uniform sphere, but that it had a very concentrated positive charge in the middle, and around that a lot of space. As atoms were electrically neutral overall, he was forced to conclude that the empty space around the centre of the atom contained negative charge that was spread out, and not so densely concentrated as the positive charge.

Rutherford is credited, as a consequence of this imaginative leap, with the discovery of the nucleus.

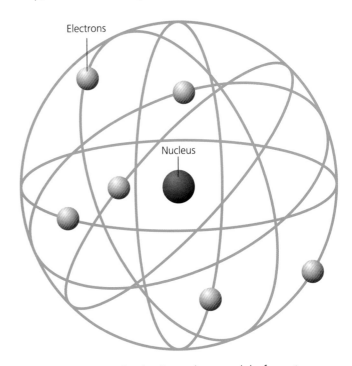

■ **Figure 11.7** Rutherford's nuclear model of an atom

■ **Figure 11.8** In reality, the nucleus is incredibly tiny relative to the size of the whole atom. If an atom were expanded to the size of a sports stadium, the nucleus would be about the size of a small marble in the middle!

Now the race was on to investigate the properties of this 'new' atom. In the space of just four decades from Rutherford's 1909 discovery, the component parts of the atom were separated out and their properties measured, and scientists had begun to manipulate the nucleus using nuclear technologies.

Because atomic particles are so small, using standard SI units makes for complicated numbers. For this reason, special atomic units are commonly used.

■ **Table 11.1** Atomic unit equivalencies

Quantity	Unit	Description	Standard SI equivalent
length	nanometre (nm)	10^{-9} of a metre	1.00×10^{-9} m
mass	atomic mass unit (u)	1/12 the mass of a carbon-12 atom, approximately the mass of 1 nucleon	1.66×10^{-27} kg
charge	electron charge (e)	equivalent charge of a single electron	1.60×10^{-19} C

■ **Table 11.2** Properties of atomic constituents

	Name	Mass (u)	Electrical charge (e)	Particle type / made from
nucleons (found inside nucleus)	proton	1.007	+1	hadrons / quarks
	neutron	1.009	0	hadrons / quarks
found outside nucleus	electron	5.489×10^{-4}	−1	lepton / elementary particle

It would be more accurate now to speak of 'models' of the atom, since one of the consequences of the breakthrough in atomic physics was the discovery that, at atomic scales, matter does not follow the same rules as those that work at larger scales. The Danish physicist Niels Bohr (1885–1962) first recognized that the idea of negative electrons orbiting the positive nucleus like tiny planets had some problems. It did not accord with experimental evidence from emission and absorption spectra in plasmas – an important discovery that we will explore further in Chapter 12. In fact, the model could only be made to work with the observations if the electrons were 'forced' to stay in orbits with certain energy states, rather like planets that can only have certain orbits. Because these energy states have a fixed amount of energy associated with them, they are said to be **quantized**.

The quantization model led to further questions about the fundamental nature of matter and a whole new branch of physics to deal with them – **quantum physics**. The quantum physics of the sub-atomic scale is very different to the 'Newtonian' mechanics of our scale, and leads to predictions and results which can be difficult to visualize. As Bohr himself is reputed to have commented, 'Anyone who is not shocked by quantum theory has not understood it.'

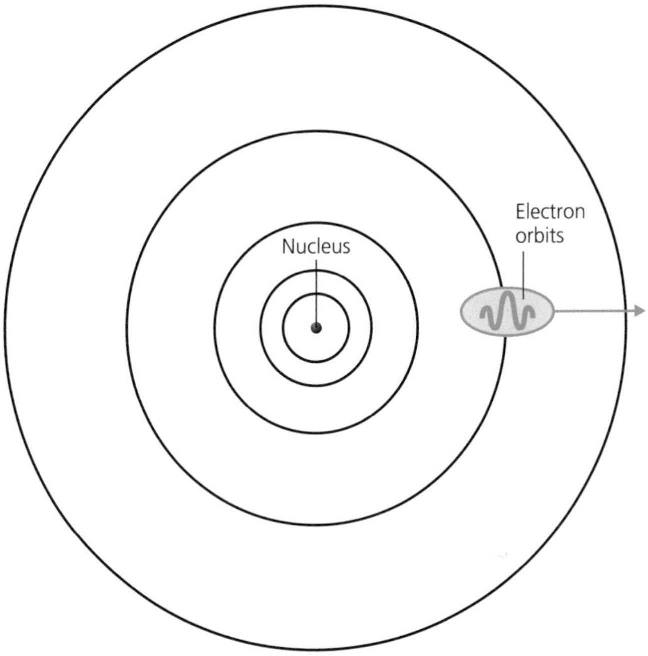

Figure 11.9 Bohr model of atom with quantized energy states

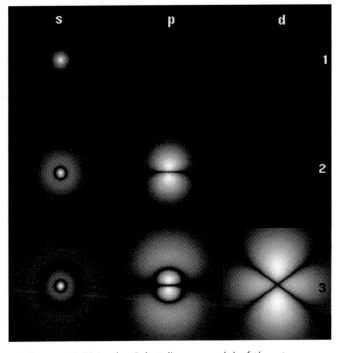

Figure 11.10 In the Schrödinger model of the atom, electrons behave like waves of whose amplitude, when squared, predicts the probability of finding an electron at that point in space

EXTENSION

If you want to find out more about quantum physics, do some research online. There are a number of helpful animations and documentary programmes available. Use wave-particle duality, uncertainty principle, de Broglie interpretation to begin your exploration.

International science

The history of atomic and nuclear physics really exemplifies the way in which scientific progress has been an international effort. While certain countries did tend to lead the way – through greater investment in research, for example – our knowledge of matter at the atomic scale is a patchwork of contributions from many different countries.

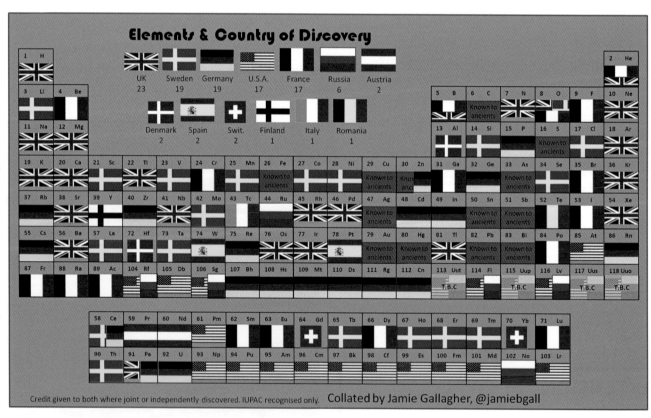

■ **Figure 11.11** A periodic table of the elements showing the countries in which the elements were discovered, courtesy of the Smithsonian Institute

Physics for the IB MYP 4&5: *by Concept*

What happens when a nucleus changes? What effects does radiation have?

ATOMS ON THE EDGE OF A BREAKDOWN

We saw in Chapter 1 how Dimitiri Mendeleev organized elements according to their properties and discovered **periodicity**. The quantized atomic model of Rutherford and Bohr made it possible to begin to explain these regular patterns of properties, in terms of the specific organization of electrons in quantized orbits. Furthermore, it was also possible to describe the elements in terms of their nuclear structure.

Figure 11.12 shows how uranium is often presented in the periodic table. The **atomic number** is equal to the number of protons in the uranium nucleus, and is sometimes called the **proton number Z**.

The **atomic mass** or **atomic weight**, tells us the relative atomic mass of the element in atomic mass units, u. It also tells us the molar mass in $g\,mol^{-1}$.

Note that the proton number tells us immediately how many electrons the neutral atom will have, since the electron has the same size electric charge as the proton, but opposite in sign, and the two must balance.

In nuclear physics it is more common to use the proton number, Z, with the **nucleon number, A**. The nucleon number is simply

the number of protons, Z + the number of neutrons, N, in the nucleus

rather than their mass as such. In nuclear physics equations, the nucleon number is written at the top left of the element symbol and the proton number at the bottom left. We would then write uranium as

$$^{238}_{92}U$$

Or, more generally for any element, X,

$$^{A}_{Z}X$$

EXTENSION

Use interactive periodic table to find online resources that allow you to explore the properties of the elements.

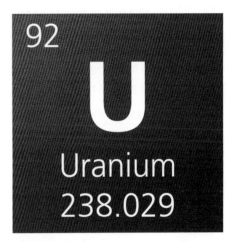

■ **Figure 11.12** Uranium, as shown in the periodic table

ACTIVITY: Charting matter

ATL

- **Information literacy skills**: Present information in a variety of formats and platforms

Use a periodic table of the elements to produce a table in a spreadsheet program of the nucleon number, *A*, and the proton number, *Z*, for as many elements as you can – but at least a minimum of 20 selected from the smallest to the largest proton numbers.

Program the spreadsheet with a formula to calculate the number of neutrons, *N*, for each of the elements. Copy the formula into your data table so that it repeats the calculation for all elements.

Now produce a **scatter graph** on your spreadsheet showing points for each of the atoms (*N*, *Z*).

	A	B	C	D
1	A	Z	N	
2	1	1	0	
3	7	3	4	
4	23	11	12	
5	39	19	20	
6	85	37	48	
7	133	55	78	
8	223	87	136	
9	9	4	5	
10	24	12	12	
11	40	20	20	
12	88	38	50	
13	137	56	81	
14	226	88	138	
15	45	21	24	
16	89	39	50	
17	175	71	104	
18	262	103	159	
19				
20				

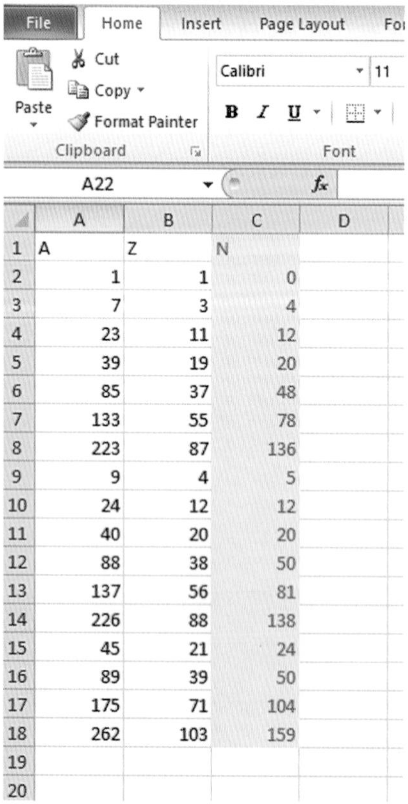

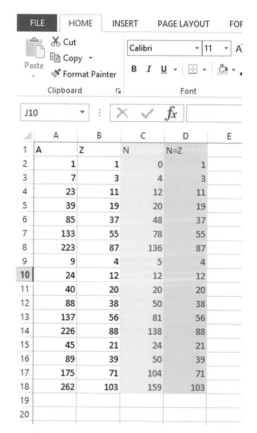

■ **Figure 11.13** Calculating neutron number *N*

Label the axes on your scatter graph clearly.

Now copy the column for *Z* again. Use this column to add a new data series to the scatter graph that shows the line *N* = *Z*.

Analysis

Outline what you notice about the curve for the actual existing elements compared with the line *N* = *Z*.

	A	B	C	D	E
1	A	Z	N	N=Z	
2	1	1	0	1	
3	7	3	4	3	
4	23	11	12	11	
5	39	19	20	19	
6	85	37	48	37	
7	133	55	78	55	
8	223	87	136	87	
9	9	4	5	4	
10	24	12	12	12	
11	40	20	20	20	
12	88	38	50	38	
13	137	56	81	56	
14	226	88	138	88	
15	45	21	24	21	
16	89	39	50	39	
17	175	71	104	71	
18	262	103	159	103	
19					
20					

■ **Figure 11.14** Comparing neutron number *N* to proton number *Z*

◆ Assessment opportunities

In this activity you have practised skills that are assessed using Criterion C: Processing and evaluating.

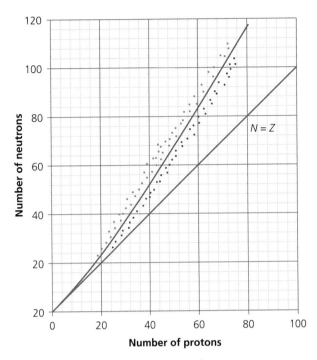

Figure 11.15 Scatter graph with $N = Z$

The curve for actual elements does not follow the line $N = Z$, but deviates upward from it at higher proton numbers. We can understand this deviation if we take a moment to apply what we already know about electrical forces to our model of the nucleus. If the nucleus consists only of protons and neutrons, then the only electrical charge present is the positive charge of the protons. While the size of the proton charge is relatively small, at around 10^{-19} coulombs, the size of a nucleus is around 10^{-15} metres, so the protons are very close together indeed. Electrostatics would predict that nuclei should, in fact, fly apart due to the electrostatic repulsion between protons – but clearly they do not!

This fact led to the hypothesis of another force acting within the nucleus over very short range. The force needs to be attractive, and it needs to 'stick' nucleons together. The **strong nuclear force** is thought to be a leftover remnant of the forces that hold tiny sub-nucleon particles called **quarks** together within individual protons and neutrons. It is a rather special kind of force – it decreases very rapidly with distance, and it only acts on certain types of particles called **hadrons**.

Since the strong nuclear force acts only over very short distances, it cannot hold very large nuclei together so well, and these nuclei, therefore, need a proportionately greater number of nucleons to provide enough force to 'glue' them together.

Figure 11.16 The Japanese physicist Hideki Yukawa (1907–1981) who first postulated the existence of sub-nucleonic particles that provided strong nuclear force

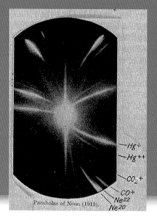

■ **Figure 11.17** Thomson's detection of neon isotopes. Thomson's note can be seen in the bottom right-hand corner of the image.

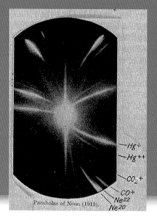

■ **Figure 11.18** Isotopes of hydrogen

At the time of writing, the properties of the strong nuclear force are still not fully understood.

Another implication of the short-range nature of the strong nuclear force is that the actual configuration of nucleons can make a difference to the stability of the atom. In 1913, the British physicist J.J. Thomson (1856–1940) was using electrical fields to deflect different kinds of atom, and he noticed while experimenting with neon that two deflections could be seen on the fluorescent screen of his apparatus.

Thomson had discovered the existence of **isotopes**. An isotope is a variation on an element: while all isotopes of an element have the same proton number (by definition), they may have different numbers of neutrons. Even hydrogen – the simplest possible element – in fact has two other known isotopes (Figure 11.18).

Some isotopes are more stable than others. The most stable forms tend to be more common, while the less stable forms may decay into other substances. In reality, whenever we gather a sample of a particular element, we will inevitably have a **mixture** of different isotopes of that element. The atomic masses in the periodic table are for most elements actually the weighted average of the masses of all the isotopes of that element.

These discoveries suggested that the nucleus was neither as solid, nor as simple as had been thought. It is the inherent instability of larger nuclei and some isotopes that is responsible for the mysterious 'rays' that Röentgen and Becquerel had noticed. We now know that these rays are, in most cases, not rays at all, but small particles of matter that are ejected from nuclei when they change to a more stable state.

Nuclei are in a constant state of energetic excitation – it is better to imagine them as wobbling, vibrating clumps of matter. Occasionally, the energy of these vibrations may overcome the short-range strong nuclear force. When this occurs, particles from the nucleus are released, carrying away some of the nucleus's energy (in fixed, quantized amounts). This process is known as **decay**. The nucleus then 'relaxes' to a lower energy state, and – if the resultant nucleus is a more stable configuration – it may even stay that way. On the other hand, there is no guarantee that the new state will be stable, so another decay may occur.

The particles of matter ejected from the nucleus are highly energetic, and electrically charged. This is why they tend to **ionize** any matter they encounter. For this reason, the mysterious rays are better termed **ionizing radiations**. This ionization can cause further changes – especially when the matter concerned is biological, in which case the radiation can cause damage to cells.

The 'alpha rays' used by Rutherford were **alpha (α) particles**. Alpha particles consist of a very stable configuration of two neutrons and two protons bound together – this corresponds, in fact, to the configuration of a helium nucleus:

$$^{4}_{2}\alpha$$

To find out what is produced when an alpha decay occurs, we have to subtract out the proton and nucleon numbers of the decaying nucleus:

$$^{A}_{Z}X \rightarrow ^{A-4}_{Z-2}Y + ^{4}_{2}\alpha$$

Note that the new nucleus Y is now a completely different element.

In addition to the kinetic energy carried by the alpha particle, some additional energy E is often released. This energy is usually very high frequency, corresponding to the gamma-ray part of the electromagnetic spectrum, and so **gamma (γ) radiation** is produced.

Another common form of ionizing radiation is called **beta (β) radiation**. This is still more exotic, since it corresponds to a fast electron that has been emitted from the nucleus. This should seem immediately counterintuitive, because after all there are no electrons in the nucleus. In fact, it turns out that neutrons can, themselves, decay to form a proton and an electron, and the excitation energy released in the process is carried away by the beta particle and another particle, called an **anti-neutrino ($\bar{\nu}$)**.

$$n^0 \rightarrow p^+ + e^-$$

$$^A_Z X \rightarrow ^{\,\,\,A}_{Z+1} Y + ^{\,\,0}_{-1} e + \bar{\nu}$$

Notice here that the nucleon number A for the new nucleus Y stays the same, since we have simply exchanged a neutron for a proton, while the proton number increases by one.

The anti-neutrino is a very exotic particle indeed – it carries no charge and has almost negligible mass, and is an example of an **anti-matter** particle. It does not usually cause any further interactions, so the effect of a beta decay is restricted to the ionization caused by the beta particles.

EXTENSION

Use alpha decay and beta decay to find examples of actual decay by elements. Check the examples using the general equations above.

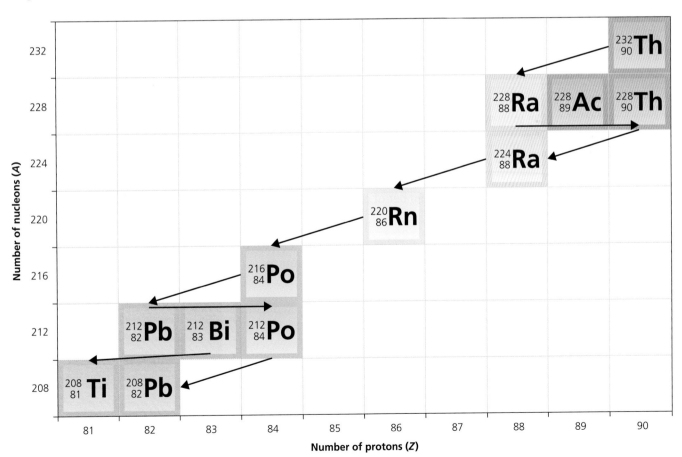

■ **Figure 11.19** Stages in the decay of thorium, showing the different elements produced

ACTIVITY: Cannonballs and mosquitos – comparing properties of radiations

Apply what you already know about atomic particle masses and electric charges to complete the first two columns of this table.

Radiation	Mass (u)	Electrical charge (e)	Penetration	Ionizing effect
alpha				high
beta				moderate
gamma				low

In this activity you will **compare** the penetration of different ionizing radiations and **explain** the results by **applying** what we have learned about their composition.

Figure 11.20 shows apparatus for comparing the penetrating effects of alpha, beta and gamma radiations.

In the experiment, three different radioactive elements are used as sources of radiation. The radiation is detected using a **Geiger–Muller tube** connected to a counter. The Geiger–Muller tube produces a current flow every time an ionizing particle enters it, and the counter counts the number of current 'pulses'.

Vibin and Kohei carry out a series of three experiments for each of three radioactive sources. The sources are
- **americium-241, $^{241}_{95}$Am**
- **strontium-90, $^{90}_{38}$Sr**
- **cobalt-60, $^{60}_{27}$Co**

In the first experiment, they measure the number of counts for 10 seconds at different distances from the detector.

In the second experiment, they measure the number of counts for 10 seconds as they insert sheets of aluminium between the detector and the source.

In the third experiment, they do the same as the second, but they use sheets of lead.

Write a hypothesis about what Vibin and Kohei should observe in their experiments. **Explain** your predictions using what you have learned about ionizing radiations.

If your school allows the use of radioactive sources, your teacher may demonstrate the experiment in action. Alternatively, your teacher will give you a table of results gathered by Vibin and Kohei.

Present Vibin and Kohei's data in a form that allows for comparison of the radioactive sources.

With reference to the data and any research of your own about the radioactive sources, **identify** the ionizing radiations produced by each of the sources and **explain** the differences in their results.

Was your **hypothesis** correct?

Evaluate the procedure in the experiment with reference to Vibin and Kohei's results. **State** any assumptions they may have made. **Suggest** how they might have improved the accuracy, or the validity of their results.

◆ Assessment opportunities

In this activity you have practised skills that are assessed using Criterion C: Processing and evaluating.

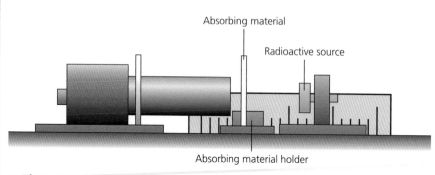
Absorbing material

Radioactive source

Absorbing material holder

■ **Figure 11.20** Set-up for radiation absorption experiment

ACTIVITY: The decay game

■ ATL

■ **Critical-thinking skills**: Use models and simulations to explore complex systems and issues

In this activity we are going to investigate radioactive decay using a simple model with dice.

Equipment
- **Minimum 30 dice or more**
- **Large container for shaking dice**
- **Pencil and paper**

Method

The dice will represent the radioactive nuclei. Choose a number that will indicate when an individual nucleus has 'decayed'.

Shake all the dice in the container.

Remove any dice that have 'decayed' – that is, fallen with your chosen number uppermost. Place these dice to one side.

Note down the number of decayed dice.

Repeat until all dice have decayed!

Results

Organize your results in a suitable table, showing number of dice remaining, turns, and number decayed.

Now **plot** your results on a graph to show how the number of dice decreased.

Analysis

On your graph, **show** the points at which the number of dice halved – whether an integer or not. So, if you began with N dice, find the points corresponding to $\frac{N}{2}, \frac{N}{4}, \frac{N}{8}$, and so on.

Now trace across to **show** the number of turns, T, at which each of these numbers was reached.

What do you notice about the intervals between each of the number of turns, T, you found?

EXTENSION

Explore further! It may be possible to obtain dice with different numbers of sides from role-play gaming shops. Repeat the game with different dice.

Outline how changing the number of sides on the dice affects the decay curve.

Perhaps you noticed in the experiment above that the experimenters started out by taking a count when no source was present – to measure the **background radiation**. While ionizing radiation can be very damaging to biological tissue, it is nevertheless present in the environment all around us. Certain rocks contain relatively high concentrations of radioactive nuclei, and some of the radiation from the Sun makes it to the ground. We are even bathed in radiation from distant stars in our own galaxy! In fact, we owe our existence to background radiation. One effect of ionizing radiation is to damage the structure of genetic molecules such as DNA. One outcome of this can be serious disease such as cancer. But over longer periods, if this damage does not destroy the organism, it can lead to **mutation** which can give rise to new life forms as natural selection takes place. Without mutation, we would not have evolved.

We have seen that, as radioactive nuclei decay, they transform into new substances which may or may not themselves be radioactive. With time, the amount of the original substance diminishes. However, we have also seen

that radioactive decay is unpredictable – there is no way to predict when the excitation energy of a particular nucleus will lead to the decay. It would seem, therefore, that it would be very hard to deduce how much of a particular substance we might expect to find in a radioactive sample.

The radioactivity of a sample of material will depend on the number of radioactive nuclei it contains. In turn, this means that the rate of decrease of the number of nuclei, N, will depend on the number, N, there are. We can write this relationship as a simple equation,

$$-\Delta N(t) \propto N$$

where the delta symbol Δ means 'change in' the number of particles, N, and (t) indicates time. (Note that the proportionality $\propto$ symbol is not the same as the symbol for alpha, α.)

This is worth a moment's pause. If the rate of decay depends on the number of nuclei remaining, what will happen to the rate as time goes on?

The decay curve (Figure 11.21) has some interesting properties.

- The gradient of the curve decreases in proportion to the number of the original nuclei, N, remaining.
- The time taken for the number of nuclei to reduce to a half of its value remains the same each time. This time is known as the **half-life** of the radioactive nucleus.

Half-life is a useful quantity, since it relates to the stability of radioactive nuclei and it tells us how much of the nucleus to expect at a given time.

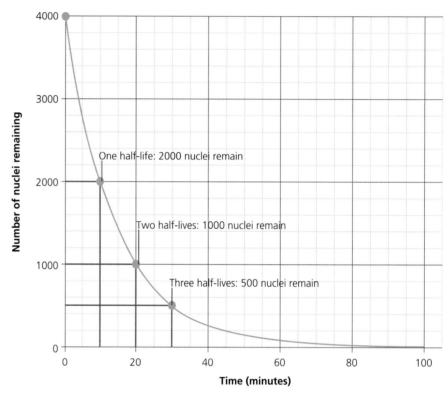

■ **Figure 11.21** Half-life decay curve

■ **Table 11.3** Sample half-lives for radionuclides

Radioactive isotope	Decay type	Half life
uranium-238	alpha	4.5×10^{19} years
uranium-235	alpha	7.1×10^{8} years
plutonium-239	alpha	24 000 years
carbon-14	beta	5730 years
caesium-137	beta	30 days
iodine-131	beta	8 days
polonium-214	alpha	0.00016 seconds

▼ Link: Mathematics

Exponentials
The radioactive decay curve is an example of a very powerful **mathematical** tool. While the term 'exponential' refers more generally to any power term in an equation, it is most often used to describe a function in which a variable's rate of change depends on the value of the variable itself. Exponentials also describe **positive feedback** processes, as discussed in Chapter 8.

ACTIVITY: Useful radiation

Radioactivity has many uses commercially, in engineering and technology, and in medicine.

Choose one of the uses of radioactivity from the list below:
- **fire and smoke detection**
- **medical imaging**
- **medical interventions – cancer treatments**
- **archaeology**
- **food preservation.**

Identify a specific technological application in your chosen area.

Describe the problem that the application solves, and **explain** how radioactivity is used in the application.

Discuss and **evaluate** the use of radioactivity for this application. Are there other ways of solving the problem? **Compare** them to the use of radioactivity, identifying important factors such as cost, health and environmental issues and cultural questions.

One very useful application of the half-life relationship is in **radiocarbon dating**. In Table 11.3 you can see that there is an isotope of carbon called carbon-14, which has a half-life of 5730 years. Carbon-14 makes up about 0.000 000 0001% of all the carbon in the environment – that is, about 1 carbon nucleus in a trillion is carbon-14! However, carbon-14 undergoes beta decay. Since it is present in the environment, it is absorbed and expelled by living things in, for example, carbon dioxide gas. While the organism is alive, the amount of carbon-14 it contains will remain about constant. However, when the organism dies, its carbon content is trapped within it. The carbon-14 component will decay. Thus, by measuring the amount of beta radiation coming from the remains of an organism – and since we know that the time for half of the carbon-14 to decay is 5730 years – we can estimate how long the living thing has been dead.

The investigation of the Turin Shroud is one famous example of the use of radiocarbon dating. The shroud is alleged to be the linen covering in which the body of Jesus Christ was wrapped after the crucifixion. Since linen is made from cotton, it could be dated. Radiocarbon dating of a tiny part of the shroud in the 1980s suggested that it was, in fact, only around 400 years old, and so probably a medieval fake. However, controversy has recently returned since it has been claimed that the linen used in the tests was a medieval patch to repair fire damage!

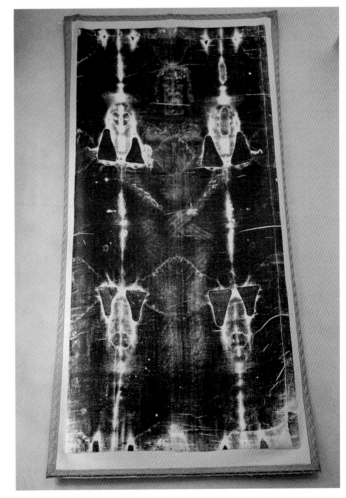

■ **Figure 11.22** Negative image of the Turin Shroud

How can we harness energy inside an atom? Is nuclear power the answer to our sustainable energy needs?

OUR NUCLEAR FUTURE?

We have seen that the nucleus, far from being a static, sold ball of matter, actually contains nucleons bound together by a short-range, strong force that counteracts the electrostatic repulsion of protons.

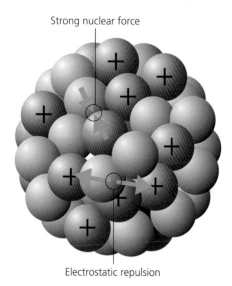

■ **Figure 11.23** Forces acting between nucleons inside an atomic nucleus

As we saw in Chapters 2 and 6, forces exist wherever there is a difference in energy. The nucleus is better visualized as an energetic bundle of particles being held together by these forces. In the nucleus, the work done in holding the nucleus together is called **binding energy**.

Large nuclei tend to be less stable than small nuclei. They require more **binding energy** overall to hold them together. The nuclei we find in nature tend to have relatively long half-lives for radioactive decay, but they can be persuaded to break into smaller pieces if we upset the delicate equilibrium of forces that holds them together.

■ **Figure 11.24** German stamp from 1979 celebrating fission

In 1934 Enrico Fermi had experimented in Rome with bombarding uranium nuclei with neutrons, and thought that he had produced elements with lower atomic numbers as a result. However, the results seemed inconclusive, and in Berlin in 1938 the German physicist Otto Hahn (1879–1968) carried out a similar experiment with his assistant Fritz Strassman (1902–1980) in which he believed he had generated barium. Hahn was not sure about his results, and consulted his colleague Lise Meitner (1878–1968).

Meitner was Jewish and had to flee the Nazi annexation of Austria for Sweden, from where she and her cousin Otto Frisch (1904–1979) corresponded by mail with Hahn. Meitner soon realized that Hahn had indeed 'split' the atom, and she calculated the process by which this could have occurred. Between them, Hahn and Meitner had discovered **nuclear fission**, the process by which large nuclei can be split into smaller ones using neutrons.

EXTENSION

Explore further! You can read Lise Meitner's original article published in the scientific journal *Nature* at www.nature.com/physics/looking-back/meitner/index.html

■ **Figure 11.25** Meitner and Hahn

In nuclear fission, a neutron is fired like a bullet into the uranium nuclei. It is important that the neutron has enough energy to enter the nucleus, but not so much that it shoots out the other side – since the nucleus would then just reassemble and no fission occurs. If the neutron is absorbed by the nucleus, on the other hand, the additional kinetic energy it introduces to the nucleus upsets the equilibrium of forces and the nucleus begins to 'wobble' or distort – Meitner compared this to the way that biological cells stretch and then split into two.

After fission of uranium, the original neutron is released, along with two further neutrons. The two smaller nuclei remaining – the **daughter nuclei** – are more stable as they require much less **binding energy** per nucleon. The excess energy is released as heat from the reaction.

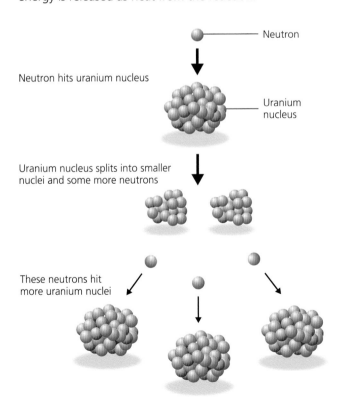

Neutron

Neutron hits uranium nucleus

Uranium nucleus

Uranium nucleus splits into smaller nuclei and some more neutrons

These neutrons hit more uranium nuclei

■ **Figure 11.26** Fission reaction

■ **Figure 11.27** A-bomb explosion

Each uranium fission releases around 3×10^{-11} J of energy. This may not seem like much, but we can calculate how much energy would be released from even a relatively small quantity of uranium metal.

1 **mole** of any substance contains the same number of particles. This number is the **Avogadro constant:**

$N_A = 6 \times 10^{23}$ atoms

1 **mole** of any substance is the mass of substance, in grams, equal to the relative atomic or molecular mass.

Atomic mass of uranium-238 = $238u$

1 mole of U-238 has a mass of 238 g. This is about the size of a grapefruit.

So 238 g of uranium contains 6×10^{23} atoms.

If all atoms undergo fission, the energy released would be

$E = 6 \times 10^{23} \times 3 \times 10^{-11} \text{J} = 2 \times 10^{13} \text{J}$

That is, around 2000 GJ of energy would be released! If all this energy is released instantaneously, the result looks something like Figure 11.27.

For this reason, research on nuclear fission was a priority for both sides in the Second World War. Ultimately, the USA was the first and only power to 'weaponize' fission in the form of two atomic bombs, which were dropped on the Japanese cities of Hiroshima and Nagasaki on August 6th, 1945.

EXTENSION

Find out more about the race to produce an atomic weapon using The Manhattan Project.

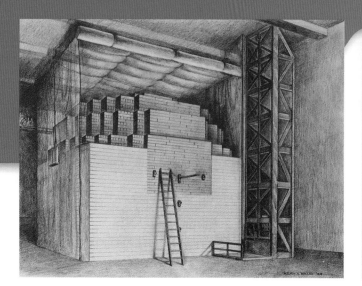

■ **Figure 11.28** Fermi's pile

■ **Figure 11.29** Interior of a nuclear reactor core being recharged with fissile fuel rods

Fortunately, it is not so easy to persuade uranium-238 to fission – or Hahn's experiment may have given unfortunate results. Although wartime certainly spurred research in atomic weapons forward, there was a lot of interest in developing **controlled** fission that might be used to produce energy. Enrico Fermi had received the Nobel Prize for his work on fission in 1938, but after receiving his prize in Stockholm he went to the United States instead of returning to Italy. Fermi had joined the Italian Fascist Party in the 1930s, but his wife was Jewish and he resigned from the party when organized persecution of Jews began in Italy. Fermi gained US citizenship and settled in New York, where he recommended work at Columbia University. Fermi was one of the chief figures in persuading the US Government to pursue atomic research. He continued work on sustainable nuclear fission, and in 1942 was moved to Chicago, where he built the first **thermonuclear pile** – under a disused swimming pool (Figure 11.28).

Table 11.4 summarizes some of the challenges, and solutions, in producing sustainable, controlled nuclear fission.

Fermi and his collaborators realized that they needed to achieve the correct balance between adequate **fissile material** to sustain fission, with a steady production of neutrons to cause ongoing fission, at the right speeds. If this could be done, the reaction would be a sustainable **chain reaction**.

When this balance is achieved, the nuclear reaction is said to be **critical**. It is – of course – equally important that the reaction does not become **supercritical**, when too many neutrons are released and an uncontrolled chain reaction leads to nuclear explosion.

Modern nuclear reactors have improved considerably on the design of Fermi's thermonuclear pile, although the underlying physics remains the same.

■ **Table 11.4**

Scientific problem	Scientific solution	Technological solution
uranium-238 does not absorb neutrons easily	isotope uranium-235 absorbs neutrons for fission more effectively, but is only 0.7% of naturally occurring uranium	separate out uranium-235 from uranium ore using gas centrifuges and use it to 'enrich' fuel to 3.5–5% uranium-235
neutrons must have correct kinetic energies to initiate fission	slow down fast neutrons to lower kinetic energies required for fission	use a **moderator** material, such as graphite, to slow down and control neutrons
too many fission neutrons can lead to a chain reaction that is uncontrollable	absorb excess neutrons	use a moderator material to absorb excess neutrons

ACTIVITY: Understanding nuclear reactors

Look at Figure 11.30. Research to find out what the different components of the design are for, and complete the table as follows:

- **In the column labelled 'Scientific problem' outline the problem that the component addresses.**
- **In the column labelled 'Technological solution' describe the way the component solves the problem.**

The first row has been completed for you.

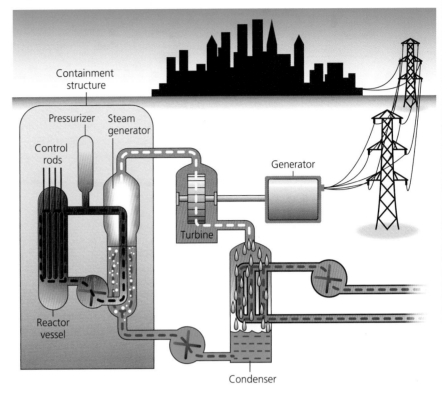

■ **Figure 11.30** Schematic diagram of a nuclear reactor

Component	Scientific problem	Technological solution
fuel rods	provide sufficient fissile material of correct type	fuel rods contain uranium oxide pellets, enriched with uranium-235
moderator		
control rods		
heat exchanger		
containment vessel		

Now, in your own words, **explain** how the design of a nuclear reactor sustains a **critical** reaction, and how it prevents a **supercritical** reaction.

■ **Figure 11.31** A nuclear reactor in Germany

Figure 11.32 France relies heavily on nuclear power as it has limited fossil-fuel reserves. Around 80% of France's electricity is generated using nuclear power

The graph in Figure 11.33 shows the **binding energy per nucleon** for elements for different **nucleon numbers**, A.

The graph reveals an unexpected result. For relatively small nuclei, the total binding energy of the nucleus is spread out over fewer nucleons. However, because the nuclei are small, less overall binding energy is required. Binding energy per nucleon increases rapidly until we get to iron, ^{56}Fe, where the curve has its peak and turning point. Iron has the most binding energy per nucleon and is the most stable element. As the nuclei get still bigger, they become less and less stable – the overall binding energy is spread out more and more and the binding energy per nucleon begins to fall away again.

We can see that, on fission of a large nucleus like uranium, we move from the right-hand side in towards the peak at iron. This shows that the binding energy per nucleon is increasing – the smaller daughter nuclei require less binding energy overall to remain stable, and energy can be released. On the other hand, to bring the daughter nuclei together to form uranium would require us to put more energy into the system.

Conversely, very small nuclei – such as the hydrogen isotopes, tritium and deuterium – require more binding energy overall than the more energy-efficient helium or nitrogen. If these nuclei were somehow put together, then the excess binding energy no longer required would be released.

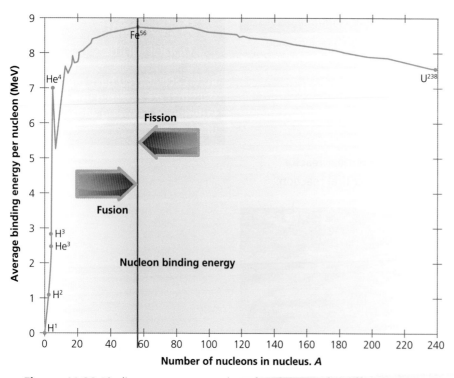

Figure 11.33 Binding energy per nucleon for elements for different nucleon numbers

The possibility of pushing small nuclei together to form larger ones was recognized by British physicist Arthur Eddington (1882–1944) as early as 1920. However, the technological barriers to achieving **nuclear fusion** were recognized to be huge. In order to bring the two small nuclei together, we must overcome the electrostatic repulsion of the nuclei. This is relatively strong, and, as we have seen, is much longer range than the strong nuclear force which might pull them together. Bringing nuclei together to fuse them requires huge kinetic energies, corresponding to temperatures upwards from $T = 10^7$ K. Eddington realized that this could happen in stars like the Sun, but not on Earth.

However, in 1952 artificial nuclear fusion was produced on Earth, in the form of a nuclear weapon. The USA detonated a test explosion equivalent to 10 million tonnes of chemical explosive, using a fission explosion to trigger fusion in liquid deuterium.

Fusion of deuterium and tritium releases nearly 2.7×10^{-12} joules per fusion – about 10 times more energy per event than fission. What is more, the by-products of the process are neutrons and helium gas. The neutrons can be absorbed and even recycled to generate more fuel for the fusion reactor, while the helium is inert and non-radioactive.

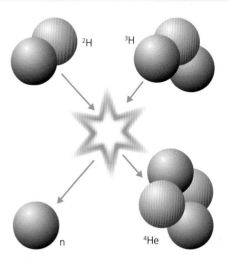

■ **Figure 11.34** Fusion of deuterium and tritium

The advantages of nuclear fusion over nuclear fission should be clear! But the technological barriers remain, after 60 years of research, unbroken.

ACTIVITY:
Understanding fusion reactors

■ ATL

- **Information literacy skills**: Make connections between various sources of information
- **Critical-thinking skills**: Analyse complex concepts and projects into their constituent parts and synthesize them to create new understanding

Research nuclear fusion using **nuclear fusion reactors**.

Use your research to complete the table, summarizing the problems of achieving nuclear fusion.

Figure 11.35 shows one possible design for a fusion reactor, called a **torus**. The design is based on an earlier Russian variant, called a **Tokamak**.

The torus uses large electromagnetic coils arranged in a doughnut shape to accelerate and contain the fission fuel as it reaches a **plasma** state.

Problem	Scientific Idea	Technological solution(s)
obtaining fuel isotopes deuterium and tritium	deuterium exists in seawater as 0.016% of hydrogen atoms	
producing high kinetic energies to initiate fusion		
removing energy from the reactor to use it		

Analyse Figure 11.35 and **outline** what you think the labelled parts do.

Use your analysis to **outline** how a torus design solves the technological problems of nuclear fusion described above.

◆ Assessment opportunities

In this activity you have practised skills that are assessed using Criterion A: Knowing and understanding.

➤

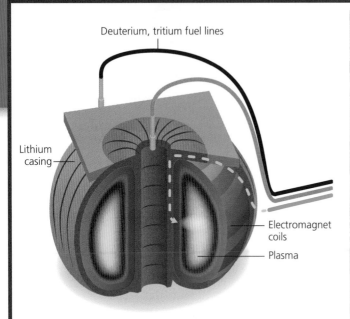

Figure 11.35 One design for a nuclear fusion reactor is a 'torus' or 'Tokamak' in whch strong electromagnetic fields are used to accelerate fusion material until it reaches a plasma state at very high energies

Deuterium, tritium fuel lines
Lithium casing
Electromagnet coils
Plasma

Figure 11.36 Interior of the experimental Joint European Torus (JET) in Culham, Oxfordshire, UK

SOURCE A

The article below is taken from a British national newspaper in 2009.

Go-ahead given for new nuclear power plants

By Pete Harrison,
Reuters – Thursday, January 10 05:50 pm

LONDON (Reuters) – Britain gave the go-ahead to a new generation of nuclear power stations on Thursday, setting no limits on nuclear expansion and adding momentum for a worldwide renaissance of atomic energy.

The government considered nuclear power unattractive as recently as 2003 but now says it will help Britain meet its climate change goals and avoid overdependence on imported energy as North Sea oil and gas supplies dwindle. Nuclear power stations provide about 18 percent of Britain's electricity now, but many are nearing the end of their lives.

'We want Britain to be more secure in its energy supply,' Prime Minister Gordon Brown told reporters. 'We are inviting companies to express an interest in building a new generation of power stations to replace the existing ones.'

Environmental group Greenpeace, which succeeded in blocking an earlier pro-nuclear decision, said the public had been misled during recent consultations and its lawyers were already considering a fresh challenge.

'This is bad news for Britain's energy security and bad news for our efforts to beat climate change,' Greenpeace Executive Director John Sauven said, adding that government plans to store highly radioactive waste underground were not safe.

SOURCE B

Read the online blog by pro-nuclear environmental campaigner Mark Lynas:

www.marklynas.org/2012/06/friends-of-the-earth-considers-abandoning-anti-nuclear-stance/

ACTIVITY: Nuclear power – yes please or no thanks

Take action

! **Take a view on nuclear power – yes please, or no thanks?**

■ ATL

- **Communication skills**: Read critically and for comprehension; Write for different purposes; Share ideas with multiple audiences using a variety of digital environments and media

- **Collaboration skills**: Manage and resolve conflict, and work collaboratively in teams; Build consensus

- **Information literacy skills**: Access information to be informed and inform others

- **Media literacy skills**: Seek a range of perspectives from multiple and varied sources

- **Critical-thinking skills**: Evaluate evidence and arguments; Consider ideas from multiple perspectives

On your own read the source materials (Sources A and B).

In pairs, **summarize** the main arguments represented in the articles.

Now research the arguments for and against nuclear power using nuclear power issues. One person can research arguments for nuclear power the other arguments against nuclear power.

Share your research with your partner.

Try to **decide** what your own opinion is on the issue.

Write the script for an online documentary about the nuclear power issue. Include a **summary** of the physics of nuclear power. Try to represent both sides of the debate equally, and **outline** arguments that are environmental, economic, and social. At the end of the documentary, **outline** your own opinion and **summarize** the arguments that support it.

Why not actually make your documentary? You can use a laptop, tablet or any other gadget with a built-in webcam. If you are good at using video editing software, you can include your own charts and diagrams.

Share your documentaries with the class, or even at a school assembly or town meeting!

◆ Assessment opportunities

This activity can be assessed using Criterion D: Reflecting on the impacts of science.

a **State** which radioactive sources most likely correspond to Type 1 and Type 2.

b With reference to the information in the table, **outline** the reason for each of the health and safety instructions.

SOME SUMMATIVE PROBLEMS TO TRY

Use these problems to apply and extend your learning in this chapter. The problems are designed so that you can evaluate your learning at different levels of achievement in Criterion A: Knowledge and understanding.

THIS PROBLEM CAN BE USED TO EVALUATE YOUR LEARNING IN CRITERION A TO LEVEL 3–4

1 Complete the table below to show the properties of the three main kinds of ionizing radiation.

Name	What is it?	Electric charge	Mass (atomic mass units)
alpha	2 protons + 2 neutrons		
beta			
gamma			

A factory uses two kinds of small, low radioactivity sources in the manufacture of smoke alarms. Workers are given these health and safety instructions:

Type 1 sources

- Always wear latex rubber gloves when handling Type 1 radioactive sources.
- Do not eat or drink when handling Type 1 radioactive sources.

Type 2 sources

- Always wear latex rubber gloves and use tongs when handling Type 2 radioactive sources.
- Do not eat or drink when handling Type 2 radioactive sources.
- Always hold Type 2 sources at least 10 cm away from the body.

THIS PROBLEM CAN BE USED TO EVALUATE YOUR LEARNING IN CRITERION A TO LEVEL 5–6

2 The graph below shows the decay of a sample of the isotope **carbon-14**.

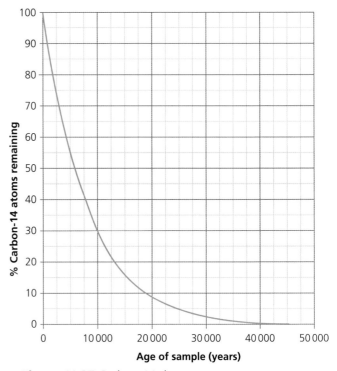

■ **Figure 11.37** Carbon-14 decay curve

a Carbon-14 undergoes beta decay. Write a beta-decay equation for carbon-14.

b Use the graph to **find** the half-life of carbon-14, in years. **Show** on the graph how you did this.

At an ancient, prehistoric burial site in the Auvergne, France, a piece of animal bone is found. Some archaeologists want to work out the age of the piece of bone.

To do this, they make some measurements which are recorded in the following table.

	Count rate (counts per second)
new bone	32
old bone	4

c Use this information and the graph to **estimate** the age of the piece of old animal bone.

d After taking the measurements, a physicist points out that the background radiation measurement was 1 count per second. **Calculate** the greatest uncertainty this would produce in the archaeologists' result for the age of the bone.

e Using your answers for **c** and **d**, **suggest** why carbon-14 dating becomes less reliable the older the sample.

THIS PROBLEM CAN BE USED TO EVALUATE YOUR LEARNING IN CRITERION A TO LEVEL 7–8

Detecting radiation

3 Scientists use different kinds of detectors to detect radiation. The kind of detector they use depends on the properties of the radiation.

Here is one kind of detector, called a **cloud chamber**:

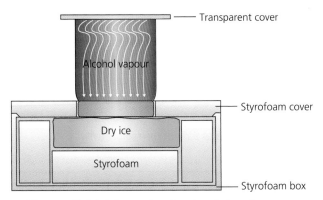

■ **Figure 11.38** Schematic of a cloud chamber

A cloud chamber consists of a closed container filled with a supersaturated vapour, e.g. water in air. When ionizing radiation passes through the vapour, it leaves a trail of charged particles (ions) that serve as condensation centres. The water vapour condenses into droplets around them. The path of the radiation is thus indicated by tracks of tiny liquid droplets in the supersaturated vapour.

■ **Figure 11.39** Photograph of ionization vapour trails inside a cloud chamber

Figure 11.39 shows what can be seen when radiation passes through the cloud chamber.

The white tracks are 'clouds' or 'con-trails' formed by the radioactive particles.

The particles are being deflected by electric fields passed through the cloud chamber.

Figure 11.40 was taken in a cloud chamber with alpha, beta and gamma radiation present. Note the direction of the electric field.

a **Identify** and **state** which 'trail' (A, B, C) was made by which type of radiation.

b **Explain** why you chose the trails that you did, with reference to the properties of ionizing radiations.

c The scientist carrying out the experiment notices that the curve of the beta track is always much 'sharper' than that of the alpha track. **Explain** why this might be.

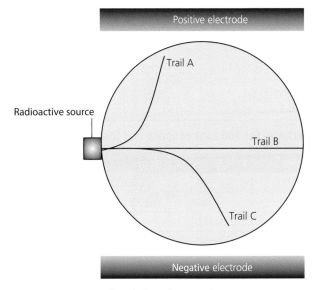

■ **Figure 11.40** Cloud chamber tracks

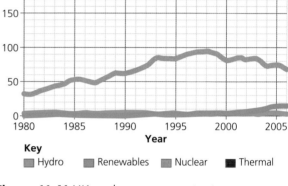

The UK's electricity production by source

Key
■ Hydro ■ Renewables ■ Nuclear ■ Thermal

■ **Figure 11.41** UK nuclear energy output

4 The table in Source 1 below shows data on different kinds of waste material left over from the core of a nuclear reactor.

a **Explain** what is meant by the word 'isotope'.
b **Evaluate** the data and **explain** which of the waste materials would be the most dangerous to handle immediately after removal from the reactor and which of the waste materials will need to be stored for the longest period of time

Read the text in Source 3 from a website.
Now **consider** the chart in Source 2.

c With reference to sources 1, 2 and 3 **evaluate** the scientific basis for the argument that the waste produced by nuclear power is not a significant problem.

SOURCE 3

www.world-nuclear.org/info/inf04.html

In the OECD countries, some 300 million tonnes of toxic wastes are produced each year, but conditioned radioactive wastes amount to only $81\,000\,m^3$ per year. In the UK for example, around $120\,000\,000\,m^3$ of waste is generated per year – the equivalent of just over 20 dustbins full for every man, woman and child. The amount of nuclear waste produced per member of the UK populations is $840\,cm^3$ (a volume of under one litre). Of this waste, 90% of the volume is only slightly radioactive and is categorized as low-level waste (with only 1% of the total radioactivity of all radioactive wastes). Intermediate-level waste makes up 7% of the volume and has 4% of the radioactivity. The most radioactive form of waste is categorized as high-level waste and whilst accounting for only 3% of the volume of all the radioactive waste produced (equating to around $25\,cm^3$ per UK citizen per year), it contains 95% of the radioactivity.

Source: accessed 18/08/10

SOURCE 1

Radioactivity for 100 tons of spent fuel (curies remaining)

Isotopes	t1/2 years	10 yrs	500 yrs	1000 yrs	10000 yrs	100000 yrs	200000 yrs
Sr-90	28	2 000 000	15	trace	-	-	-
Cs-137	30	3 000 000	40	trace	-	-	-
Pu-239	24 110	22 000	27 000	29 000	56 000	8000	240
Pu-240	6540	49 000	175 000	170 000	69 000	7	trace

*A typical 1000 megawatt reactor contains about 100 tons of enriched uranium, one-third of which is renewed each year.

Taken from Warf, James C., All Things Nuclear, First Edition, Southern California Federation of Scientists, Los Angeles, 1989, p.85.

Reflection

In this chapter we have explored the physics of the atom's interior, to understand how the nucleus is structured and what its component parts are. We have characterized the properties of different ionizing radiations, and analysed what happens in the nucleus when they are released. We have then explored the technologies humanity has devised to control and utilize nuclear reactions to produce energy.

While nuclear power has become well established in the past 50 years, it remains controversial with many environmental campaigners.

Use this table to reflect on your own learning in this chapter.					
Questions we asked	**Answers we found**		**Any further questions now?**		
Factual: What's inside an atom? What happens when a nucleus changes? What effects does radiation have? How can we harness energy inside an atom?					
Conceptual: What new possibilities does nuclear technology offer? What special consequences does nuclear technology bring?					
Debatable: Is nuclear power the answer to our sustainable energy needs?					
Approaches to learning you used in this chapter:	**Description – what new skills did you learn?**	**How well did you master the skills?**			
		Novice	Learner	Practitioner	Expert
Critical-thinking skills					
Information literacy skills					
Media literacy skills					
Affective skills					
Communication skills					
Collaboration skills					
Learner profile attribute(s)	Reflect on the importance of an open-minded attitude for our learning in this chapter.				
Open-minded					

12 Where are we in the Universe?

As we extend the reach of our **observations**, we better understand the **relationships** that **form** our **models** of the Universe, and so our **place in the cosmos**.

■ **Figure 12.1** Earth from space

CONSIDER AND ANSWER THESE QUESTIONS:

Factual: What is the scale of the observable Universe and how big are the objects in it? What evidence have we used to elaborate our models of the Universe? What instruments have we used to gather observational evidence? How do stars produce energy?

Conceptual: Why is the speed of light important to our understanding of the Universe? How do forces shape the Universe?

Debatable: How has knowledge of the Universe affected our understanding of our place in it? How important is it to know about the Universe beyond our own planet?

Now **share and compare** your thoughts and ideas with your partner, or with the whole class.

IN THIS CHAPTER, WE WILL ...

- **Find out** how the earliest space scientists explored the Solar System and the objects in it, and about the optical instruments such as the human eye, lenses and telescopes that are used to manipulate light and achieve magnification of distant objects.
- **Explore** other forms of observational evidence available to astronomers as they looked further out into deep space, such as spectroscopy; the implications of a finite speed of light, and of the expanding Universe, for our own understanding of our place in space and time.
- **Take action** to evaluate the benefits and costs of space research.

■ These Approaches to Learning (ATL) skills will be useful ...

- **Information literacy skills**
- **Critical-thinking skills**
- **Transfer skills**

● We will reflect on this learner profile attribute ...

- Knowledgeable – in this chapter we will explore how our knowledge of our place in the Universe has grown.

■ **Figure 12.2** Small blue dot

◆ Assessment opportunities
   in this chapter

◆ **Criterion A**: Knowledge and understanding
◆ **Criterion C**: Processing and evaluating
◆ **Criterion D**: Reflection on the impacts of science

KEY WORDS

galaxy	moon	Sun
image	planet	Universe
magnify	star	

SEE-THINK-WONDER

Look at Figures 12.1 and 12.2.

The first is a composite image produced by NASA, showing the Earth as seen from orbit.

The second picture was taken by *Voyager 1* in 1991 as it approached the outer limits of our Solar System. It also shows the Earth, as indicated by the arrow.

What do these pictures make you think?

What do these pictures make you wonder?

When we look up on a clear night, we will see the stars. This is true no matter where we are on Earth (although the stars we see might be different). Reflecting on our place in the Universe can bring humans together across time, and across space.

We began our physics programme by thinking about how things looked at different scales. In Chapter 11 we zoomed in to the smallest possible scales, inside the atom. In this final chapter we zoom out to the biggest of scales and explore the physics that has enriched our understanding of our place in the Universe.

What is the scale of the observable Universe and how big are the objects in it?

WHAT IS OUT THERE?

Stars are not the only things we can see when we look up in the night sky. Figure 12.3 shows part of the night sky taken from the Northern hemisphere over a few minutes.

■ **Figure 12.3** Streak formed by the ISS in long-exposure photograph of the night sky

You can see the stars in the image as clear, still points of light. But the white streak is caused by something much closer to home – it is the light trail left by the International Space Station (ISS) as it moves in orbit over the Earth. Similarly, at different times of the year and from different parts of the Earth, we can see our nearest neighbours, the planets, moving through the night as Earth turns on its axis.

As we saw in Chapter 2, the most influential force acting on masses over long distances in space is gravity or **gravitation**. It is gravity that determines the fundamental forms and geometry of the Universe, whether way out in deep space or in our own spatial neighbourhood. The most important gravitational object near to us is, of course, our Sun, and the Sun's gravity is the most significant influence in our own **Solar System**. The Solar System consists of

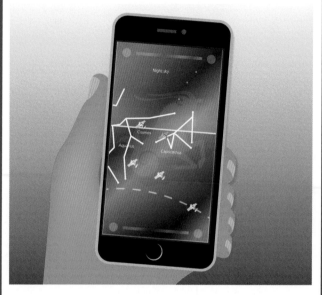
many different kinds of objects, but all of them are bound into **orbits** by the Sun's gravitation. The objects themselves all exert gravitational force on each other, and this can lead to changes or **perturbations** in their orbits – Jupiter, the most massive of the planets, is particularly important in this respect.

ACTIVITY: Sorting the Solar System

- **Information literacy skills**: Access information to be informed and inform others
- **Critical-thinking skills**: Gather and organize relevant information to formulate an argument

Figure 12.5 shows the principal objects in the Solar System. The labels in the box are mixed up.

In pairs, discuss which label goes with which object.

| inner planets | asteroid belt | outer planets |
| satellites | comets | oort cloud |

Research the categories in the table and then **classify** the objects using the categories. **Outline** the characteristics of each of the categories using examples from the Solar System schematic.

	Examples	Approximate diameters (m)	Characteristics
planetoids		$1-10^3$	
satellites/ moons		10^4-10^6	
rocky planets		10^6-10^7	
gas giant planets		10^8	
stars		10^9	

Now **research** to find accurate values and so **estimate**:

- the number of rocky planets like the Earth that would fit across the diameter of Jupiter
- the number of rocky planets like the Earth that would fit across the diameter of the Sun.

■ **Figure 12.5** Schematic diagram of the Solar System

Just as for the tiny scales of atomic physics, it can be difficult for us to visualize the scales of space physics. Astrophysicists have defined some different 'rulers' to make the scale of things easier to manage. Within the Solar System, one such 'ruler' is the **astronomical unit**:

1 astronomical unit (AU) = mean distance from centre of Earth to centre of Sun = 149.6 million km

Our understanding of the form of the Universe, and of our place in it, has changed as the nature and quality of our observations have changed. In the past, many people assumed that the Earth was flat, since it just looked that way. Eratosthenes (276–194 BCE) estimated the circumference of the Earth using the elevation of the Sun at noon in Alexandria and in Syene (now Aswan), and in 1492 Christopher Columbus used his estimation to justify his voyage to Asia – accidentally running into the 'New World' of the Americas instead. Similarly, for a long time, it was generally held in Europe that the Earth was at the centre of the Universe, with planets and the Sun orbiting around it. This **geocentric** view derived from the astrological cosmology of Ptolemy of Alexandria (died 168 CE). It is interesting that many people assume this was the 'natural' thing to think – after all, it certainly **looks** like objects such as the Sun and the Moon move around us as we stand still. However, as the philosopher Ludwig Wittgenstein is reputed to have once remarked at a dinner party, 'How would things look if, in fact, the Earth moved around the Sun?' Wittgenstein's point was that in actual fact we **are** moving around the Sun – the observations alone cannot explain why we would assume that Earth was at the centre of everything! So, why would we assume that our Earth was at the centre of the Universe?

Appearances can be deceptive

In the cases of the geocentric and the flat-Earth cosmological models, first appearances can be deceptive. While science relies on observations to make preliminary hypotheses about the Universe, not any old observation will do: the observations we make have, themselves, to be selected so as to suggest a relationship between key variables. The relevant variables are chosen through having a 'working idea' – or hypothesis – about the phenomenon to be investigated. Some philosophers have argued that a scientist's first 'working idea' is really no more than an imaginative guess!

By the time of Galileo Galilei's birth in 1564, the idea that the Earth was spherical was well established in Europe, but the idea that the Sun was at the centre of things was still controversial. The claim had been made by a Polish astronomer, Nicolaus Copernicus (1473–1543) and his extensive calculations based on relatively accurate observations of the motion of planets were persuasive. Some sections of the Roman Catholic Church at the time thought that the geocentric view was supported by Biblical text – although the Pope had listened with interest to a description of the **heliocentric** model as early as 1533, and in 1588 the heliocentric system was used by the Church to make corrections to its own calendar.

In the early 1600s, Galileo was a professor at the University of Padua and he made a telescope based on rumours he had heard from Holland of such a device. Galileo did a good trade in selling telescopes to Venetian merchants as a way to spot pirates at sea, but he also turned his new instrument to the skies. During the period 1609–1610 Galileo made incredible new observations of the Moon and the planets, especially the planet Jupiter (see Chapter 1). Galileo's observations of Jupiter showed four points of light moving from one night to the next, and he realized that he was observing a system of moons orbiting the planet. This convinced him that Copernicus has been right – surely, if Jupiter were the centre of the 'cosmos' for these moons, then the Sun could be at the centre of ours?

Unfortunately Galileo was never very careful about the way he made his points. He published his arguments in favour of the **heliocentric** system in a book called 'The Starry Messenger' in 1610, and these ideas were used by his enemies to have Galileo condemned for **heresy**. Faced with the choice of recanting (retracting), or of being burned at the stake, Galileo chose to retract.

All the same, the idea was now well and truly out, and Galileo's reputation was strong across Renaissance Europe. Johannes Kepler (1571–1630) refined the heliocentric model still further, using careful observations made by the astronomer to the Danish royal court, Tycho Brahe (1546–1601), and Kepler elaborated three important laws on the basis of his new calculations.

■ **Figure 12.6** Johannes Kepler

Kepler's three laws of planetary motion

1. The law of orbits: all planets move in elliptical orbits, with the Sun at one focus.
2. The law of areas: a line that connects a planet to the Sun sweeps out equal areas in equal times.
3. The law of periods: The square of the period of any planet is proportional to the cube of the semi-major axis of its orbit.

In the geocentric view, the Sun and the planets had been held in place by slowly rotating crystal spheres arranged concentrically around the Earth – now Kepler's realization that the orbits of the planets were not spherical but **elliptical** meant this idea would not work. So what was holding the planets in their orbits?

GRAVITATION

Kepler's third law is a consequence of the second – the further out we go, the slower a planet's orbit. In 1684, nearly a hundred years after Kepler published his laws, Isaac Newton suggested that the 'something' that held planets in their stately orbits was a force, which he named **gravitation**.

This 'gravitation' was hard to accept. It was unseen, but its effects were indisputable – you may have heard the story about Newton's inspiration when he was woken from a nap under an apple tree by a gravitationally influenced apple. What was more, gravtitation acted between all masses, and everywhere. We have already encountered the outline form of gravitational force fields in Chapter 2, and Newton used Kepler's work to formulate his law of universal gravitation:

$$F = G\, \frac{m_1\, m_2}{r^2}$$

where m_1 and m_2 (kg) are two masses exerting gravitational force on each other, at a distance r (m) apart. G is the universal gravitational constant, and its value was first measured by the British physicist Henry Cavendish in 1798. The best modern value for this constant is 0.000 000 000 0667 newton metre squared per kilogram squared ($6.67 \times 10^{-11}\,\mathrm{N\,m^2\,kg^{-2}}$).

ACTIVITY:
Investigating Kepler's laws

■ ATL

- **Critical-thinking skills**: Use models and simulations to explore complex systems and issues

Draw a diagram to illustrate Kepler's first law.

Now **demonstrate** Kepler's second law for Earth, Mars and Jupiter. Refer to the data in the table below.

Planet	Mean orbital radius (AU)	Orbital time period (years)
Earth	1.0	1.00
Mars	1.5	1.88
Jupiter	5.2	11.9

On a large sheet of millimetric paper and using a scale of 2 cm = 1AU, **draw** circular orbits for each of the three planets.

Calculate the angle the Earth will move through in 3 months.

Draw and shade a segment of Earth's orbit with this angle at the centre

Now **calculate** what angle Mars and Jupiter will move through in the same time, and **draw** and shade similar segments.

State what Kepler's second law says about the area of the segments you have drawn.

Suggest what this tells us about the speed of the planets in their orbits.

By **labelling** your diagrams with suitable variables, write Kepler's third law in the form of an equation.

Assessment opportunities

In this activity you have practised skills that are assessed using Criterion A: Knowing and understanding.

ACTIVITY: Massive force

ATL

- **Critical-thinking skills:** Draw reasonable conclusions and generalizations

Look at the table showing data in the Earth–Moon system.

Mean Earth–Moon distance (m)	3.8×10^8
Mass Earth (kg)	7.3×10^{22}
Mass Moon (kg)	5.9×10^{24}

Use this data on planetary masses, some suitable **estimates** and Newton's law of universal gravitation to calculate:

- **the gravitational force of the Moon on the Earth**
- **the gravitational force of the Earth on the Moon**
- **the gravitational force of your physics teacher on you.**

Comment on the relative sizes of these forces.

Newton used Kepler's earlier work to devise his law. **Suggest** how Newton's law agrees with Kepler's observations.

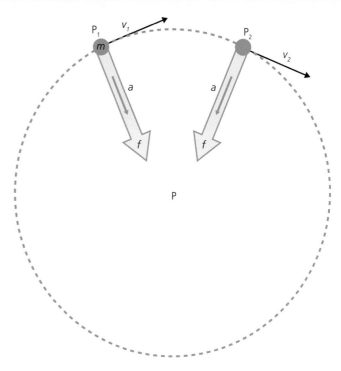

■ **Figure 12.7** In the diagram the mass m is travelling in a circle around point P. In position 1, P_1, the mass has a velocity v_1 at a tangent to the circle. At position 2, P_2, the mass has velocity v_2. While the rotational **speed** is the same at these points, the velocity has changed direction towards the centre of the circle. This means there is a centripetal acceleration a and a centripetal force f acting towards the centre of the circle

Newton's law of universal gravitation follows as a consequence of the second law of motion. Any object moving in a circle is changing direction constantly. As we saw in Chapter 5, a change in direction means a change in **velocity**, and this, in turn, means an **acceleration** must take place. While the **inertia** of the object means it would usually continue in a straight line, instead the object is 'moving' as though it were trying to reach the centre of the circle, so the acceleration is directed **inwards**.

Newton's second law

$$\vec{F} = m\vec{a}$$

tells us that a force is required to produce this acceleration, and the force is in the same direction: inwards towards the centre of the circle. This force is called **centripetal force**.

When an object orbits around a gravitational mass, such as the Earth, it is pulled towards the centre of the Earth by the centripetal force provided by its weight. If the object is moving quickly enough, the Earth's surface will curve away from it by a distance equal to its fall, and so the object effectively stays in an **orbit** in a state known as **freefall** (Figure 12.8).

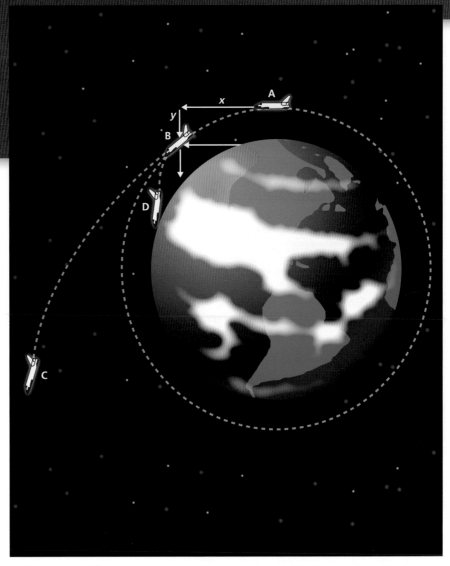

■ **Figure 12.8** The spacecraft stays in a circular orbit between A and B. In travelling the tangential distance *x* it falls a distance *y* towards the Earth surface, but the Earth's surface curves away from the spacecraft at an equal distance *y*. To reach C, the spacecraft needs to increase its tangential velocity and leave circular orbit. To land via D, the spacecraft would decrease its tangential velocity

EXTENSION

The Earth is ringed by many thousands of artificial satellites that we have launched over the period since *Sputnik 1*, the first human-made satellite sent into space by the Soviet Union in 1957. Find out about how satellites in different kinds of orbits are used for different purposes using these search terms:

- polar orbit
- geostationary orbit
- global positioning network
- meteorological satellite
- communications satellite
- spy satellite.

THE LONGEST VIEW

Kepler relied on Brahe's observations to work out the detail of his laws of orbital motion – and Brahe had only his own eyes, and some geometrical instruments he designed himself, to make his observations. The most important instrument we have is our own eye, and it is highly sophisticated. To understand how the eye works, we need to return to some of the properties of light we examined in Chapter 7.

When an object is illuminated, light waves undergo **reflection** from it in all directions. We saw in Chapter 7 how those waves spread out from points on the object, and we indicated the direction of wave travel with **rays**. In order to 'see' an object, we need to collect rays from all points on the object together, such that the rays are all organized correctly. These rays will then produce an **image** of the object.

In Chapter 7, we also saw how transparent materials cause **refraction** of incident light, as well as reflection. A **lens** is a special kind of **prism**, whose shape is designed so as to organize the incident rays in a particular way.

■ A **biconvex** lens has both sides curved outward towards the middle. This has the effect of refracting rays that are parallel to the **axis** of the lens **inwards**, so that they cross at a **focal point**.

■ A **biconcave** lens has both sides curved inward towards the middle. This has the effect of refracting rays that are parallel to the **axis** of the lens **outwards**, as though they had all originated at a **focal point** behind the lens.

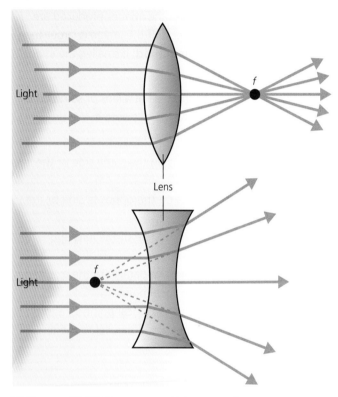

■ **Figure 12.10** Biconvex and biconcave lenses

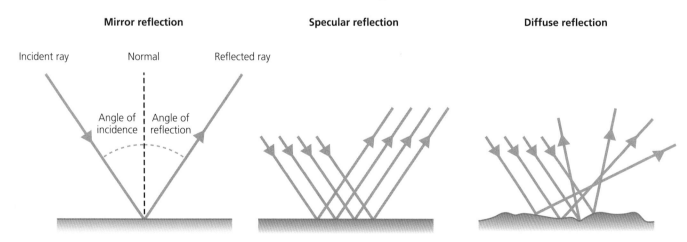

■ **Figure 12.9** Types of reflection

In the Activity: Making images, the images produced are different sizes to the real objects. If the image is larger than the object, it is said to be **magnified**; if smaller, the image is said to be **diminished**.

A **ray diagram** is used to show what a lens or some other optical device does to light which falls onto it. We usually draw three rays and two focal points.

- The **near focal point**, f_1, is on the same side of the lens as the object.
- The **far focal point**, f_2, is on the opposite side of the lens to the object.
- One ray goes through the middle of the lens and is not refracted.
- One ray goes through the lens parallel to the axis and is refracted through the far focal point, f_2.
- One ray goes through the near focal point, f_1, and then through the lens and is refracted parallel to the axis.

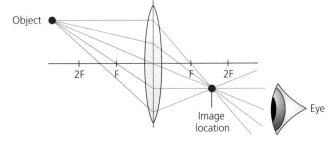

Figure 12.11 Ray diagram – biconvex, real 1

ACTIVITY: Making images

ACTIVITY: Making images

■ ATL

- **Critical-thinking skills**: Practise observing carefully in order to recognize problems

Aim: To observe the effect of biconvex and biconcave lenses.

Equipment
- **Biconvex lens**
- **Biconcave lens**
- **Bright source of light**
- **Piece of squared paper**

Experiment 1

Hold the biconvex lens up to a bright source of light that is a fairly large distance away, for example on the other side of the room (a window, or a light bulb).

Hold a piece of squared white paper behind the lens and move it steadily to and fro until you see an image produced on it.

Describe the image you see with some of the words from the box.

diminished	inverted	magnified	upright

This image is being projected into an actual, real position in space: this is why we can see it on the screen. For this reason, it is called a **real image**.

Now try the same with the biconcave lens. **Describe** what you see on the paper.

Experiment 2

Take the biconvex lens and hold it close over this page. Look through the lens. Move the lens to and fro until you see a clear image.

Describe what you see in the lens, using the correct terms from the box.

This image cannot be projected onto a screen, because it does not exist as such in real space: it is being produced by the rays that pass through the lens and into your eye. This is called a **virtual image**.

Now try the same with the biconcave lens. **Describe** what you see this time.

In Figure 12.11, we can see that the rays which come from the top of the object are all reassembled on the other side of the lens at the same point, relative to the lens. This forms a **real image**, as we see in the Activity: Making images. We can generalize to rays from any point on the object, and see that they are all reassembled at the corresponding point in the image (Figure 12.12).

Ray diagrams are useful schematic simplifications because they allow us to model the behaviour of light passing through the lens, and they give us information about the kind of image that will be produced.

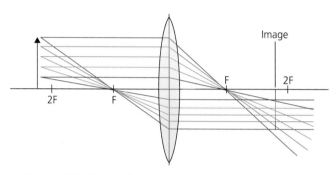

■ **Figure 12.12** Ray diagram – biconvex, real 2

On the other hand, when we hold the lens over an object that is closer to it than the focal distance, f, and we look **through** the lens, something quite different happens.

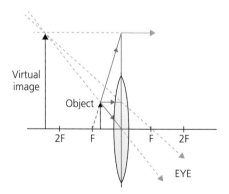

■ **Figure 12.13** Biconvex, virtual

ACTIVITY: Constructing ray diagram images

■ ATL

■ **Critical-thinking skills**: Use models and simulations to explore complex systems and issues

Construct ray diagrams for the following situations:
● **Object between focal point f and a point $2f$ from the lens.**
● **Object at point $2f$ from lens.**
● **Object beyond point $2f$ from lens.**
● **Object placed at the focal point f.**

Describe the images formed in each case using the words in the box from the previous activity (page 248).

■ One ray from the object will pass unrefracted through the middle of the lens.
■ One ray will travel parallel to the axis through the lens, and then be refracted through the far focal point, f_2.
■ One ray will appear to have come from the near focal point, f_1, and be refracted parallel to the axis.

If we extend these three rays, we can see that they now **diverge**. So how can they form an image? If we place our eye in the position shown (Figure 12.13), on the far side of the lens, our eye reassembles the light rays so that they **appear** to have originated from an image on the near side of the lens. This image does not 'exist' in real space, but is a construct of our eye and our brain – so it is called a **virtual image**.

Our eye contains a lens system, formed by the **cornea** and the **lens** (Figure 12.14). The cornea has a **refractive index** of around 1.4 and collects much of the light that will form the image in the eye, but the lens can be stretched by the **ciliary muscles** to change the focal length of the system a little – this allows the eye to **accommodate** for near and far objects. The iris can also change size, in order to control the amount of light entering the eye – too much and we are blinded, too little and we cannot see at all.

The retina is covered with **photosensitive** cells that turn light waves into electric impulses which are transmitted by the **optic nerve** to the brain. It is the brain, in the end, which makes sense of the information so that we can 'see'.

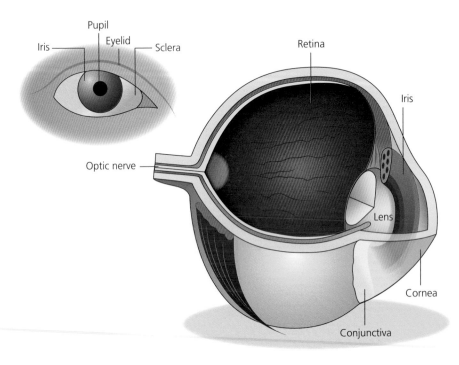

■ **Figure 12.14** Schematic parts of the eye

ACTIVITY: Eye test!

■ **ATL**

■ **Critical-thinking skills**: Practise observing carefully in order to recognize problems

We can experience the function of the different parts of the eye ourselves. Try these different tests with a partner.

Eye test 1

One person (the subject) closes their eyes and covers them with their hands, so as to block out as much incident light as possible. Remain like this for two minutes.

The other person (the observer) sets up a bright, illuminated object, such as the light bulb from the previous lenses activity (page 248).

The observer then positions themselves so as to be able to see into the subject's eyes (when they open them).

The subject then opens their eyes and looks at the bright object. The observer should watch the effect on the subject's iris.

Eye test 2

The subject holds a sheet of printed text at the end of their nose. Then, slowly moves the sheet backwards until they can just begin to read it in focus. The observer measures the distance of the sheet from the subject's eyes. This distance is called the **near point**.

Next, the subject looks up from the sheet to some text on a poster some distance away. They should tell the observer when they can first begin to read this text in focus. The observer measures this time.

Conclusions

Using what you have learned about lenses and the eye, **outline** what is happening in each of the eye tests above.

ACTIVITY: Artificial eyes?

■ **ATL**

■ **Transfer skills**: Apply skills and knowledge in unfamiliar situations

A camera functions, in many ways, like an artificial eye.

Use camera parts functions to find out what the following parts of a camera do. **Compare** to the human eye, and suggest what part of the eye fulfils an equivalent function.

Camera part	Function	Equivalent part of eye
shutter		
diaphragm		
lens		
film or charge-coupled device (CCD) photosensor		

Galileo's version of the telescope used two biconvex lenses together. In a telescope, the first lens or **objective** lens is used to gather the light from a distant object, and so it forms a **real image** inside the telescope tube. This image then forms the object for the second or **eyepiece** lens. The eyepiece is adjusted so that the real image is close to the lens and so it magnifies the real image again, forming a much **magnified virtual image** in the eye of the observer. It is thought that Galileo's telescope achieved a magnification around ×3, although even small modern telescopes achieve much bigger magnifications than this (Figure 12.15)!

Another way to produce a large magnification is to use a curved mirror as the objective. A concave mirror has the effect of focussing light rays in the same way as a convex lens. Newton developed a telescope that used a concave mirror to produce a real image that was then magnified by an eyepiece lens in the same way as the astronomical refracting telescope (Figure 12.16).

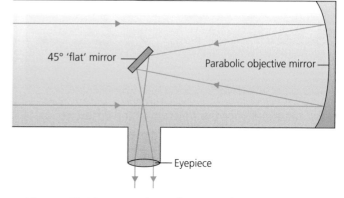

■ **Figure 12.16** Newtonian reflecting telescope

The reflecting design has a number of advantages over the refracting (lens) telescope.

- It is much easier to make high-quality, accurate mirrors with precise focal points than it is to make accurate lenses.
- Mirrors reflect all wavelengths in the same way. Lenses refract different wavelengths by different angles.
- Reflecting telescopes can be shorter than refractors.

Most astronomical telescopes – although not all – are now reflector telescopes of different types. They produce extremely high magnification and **resolution**.

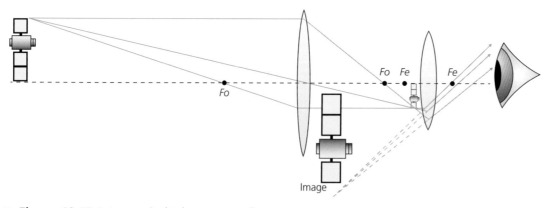

■ **Figure 12.15** Astronomical telescope ray diagram

Figure 12.17 The Keck telescopes at Mauna Kea observatory, Hawaii, use 10 m diameter mirrors with 17.5 m focal lengths

Astronomical observatories are generally situated on mountaintops, high above sea level. As we saw in Chapter 8, the atmosphere acts as a very effective radiation filter. While this is very good for life on the Earth's surface, it is not so good for astronomers who want to gather as much information as they can from the radiation arriving at the Earth.

The highest telescopes are not on Earth at all, but in orbit around Earth. The Hubble Space Telescope was one such astronomy platform launched in 1990, carrying a 2.4 m diameter optical reflecting telescope.

EXTENSION

Explore further! Since its launch, the Hubble has produced astonishing images of deep-space objects. Use NASA's dedicated site http://hubblesite.org/ to explore the Universe through the eye of Hubble.

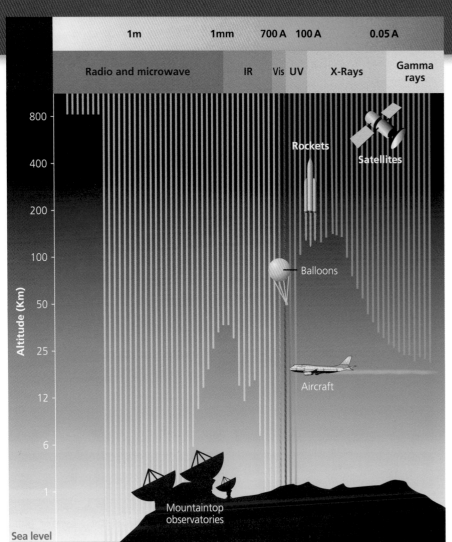

Figure 12.18 Radiation penetration of atmosphere

Figure 12.19 Hubble space telescope

But seeing is only part of the story. While optical telescopes give us valuable information, the Universe is alive with radiation from right across the **electromagnetic spectrum** (see Chapter 7). Astronomers use telescopes that are designed to detect radiation from the radio right through to the gamma ray regions of the spectrum, allowing us to 'see' the Universe through very different eyes, each revealing different information to us.

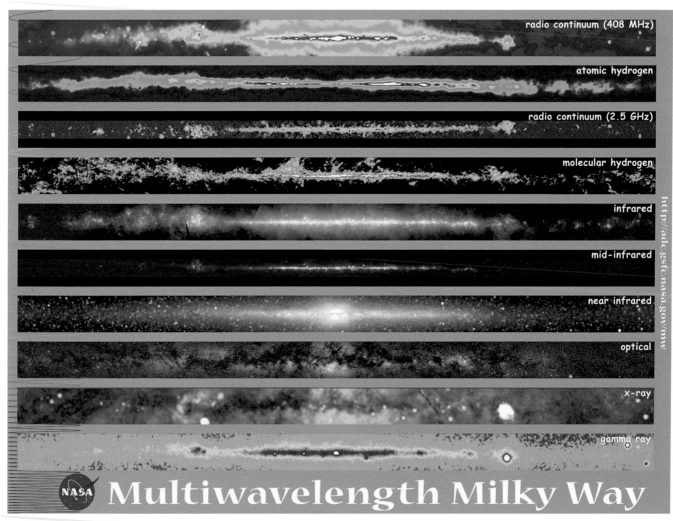

■ **Figure 12.20** Photomontage of galaxies/different wavelength regimes

How do stars produce energy?

THE LIFE OF A STAR

People say 'as sure as night follows day' to show how certain something is. There seems nothing more constant and certain than the Sun rising and falling each day. The stars, too, seem the same each night, constant and still. In Europe, ancient people thought that the stars were 'eternal', stationary in the sky and never changing. This idea was reflected in Ptolemy's geocentric model. However, in a Chinese book called *Ch'ien-han-shu*, recording astrological events back to the year 5 BCE, historians have found the reference:

> In the second year of the period of Ch'ien-p'ing, second month, a hui-hsing appeared in Ch'ien-niu for more than days.

Similarly in the Korean chronicles *History of the Three Kingdoms – The Chronicle of Silla (Samguk Sagi)*, a reference is found to:

> Year 54 of Hyokkose Wang, second month Chi-yu, a po-hsing appeared in Ho-Ku.

The Chinese and Korean astrologers recorded a bright 'guest star' dating in modern calendar terms to July 1054, in a place where we now see the Crab Nebula. Then again, in 1572 Tycho Brahe made brightness measurements of a star which suddenly lit up the whole sky at night, then disappeared after a few days.

When we make observations of the Universe around us, we see stars of many different kinds to our own Sun. Astronomers have discovered that a star does indeed change quite dramatically during its 'lifetime'.

Figure 12.21 Crab nebula supernova remnant

ACTIVITY:
It's a star's life

Use life cycle star to research using online videos about the life cycle of stars.

Match the labels and descriptions to the parts of the schematic diagram (Figure 12.22).

Interpret the information you have collated in the diagram to **deduce** the variable that determines whether a star will form a black dwarf or a black hole.

Label	Description
stellar nebula	large red star formed when a massive star begins to fuse heavy elements in its core
neutron star	very massive object whose gravitational pull prevents even light from escaping
massive star	small, bright star that is fusing heavy elements very rapidly
red supergiant	large cloud of gas and dust slowly collapsing under gravitation
main sequence star	very dense remnant of a supernova where individual nucleons have fused under gravitation to form neutrons
planetary nebula	very large stars that are rapidly fusing hydrogen to helium and heavier elements still
black dwarf	star around the same size as the Sun that is mostly fusing hydrogen to form helium
white dwarf	gaseous remains of outer layers of a main sequence star blown into space
red giant	'dead' remnant of a white dwarf made mostly from iron
black hole	exploding core of a massive star after supergiant collapses inward, very bright for a short time
supernova	large red star formed when a main sequence star begins to fuse heavy elements in its core

◆ Assessment opportunities

In this activity you have practised skills that are assessed using Criterion A: Knowledge and understanding.

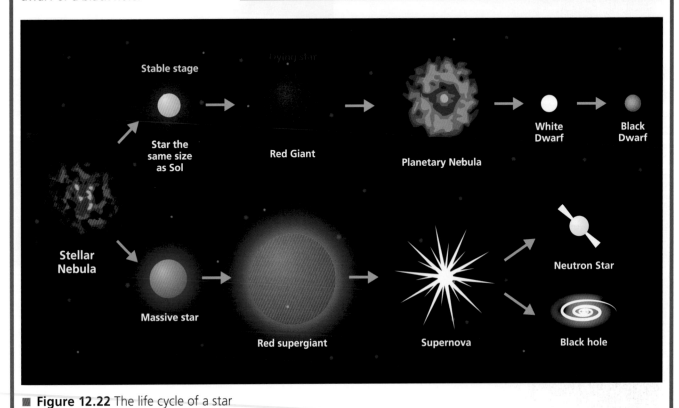

Stable stage

Star the same size as Sol

Red Giant

Planetary Nebula

White Dwarf

Black Dwarf

Stellar Nebula

Massive star

Red supergiant

Supernova

Neutron Star

Black hole

■ **Figure 12.22** The life cycle of a star

The time taken for a star to complete its life cycle depends on its mass, but it is always tens of billions of years. Fortunately for us, our Sun is a fairly average 'main sequence' star in the middle of its life, about 4.6 billion years through its likely lifetime of 10 billion years. The Sun is a huge sphere, largely consisting of hydrogen gas. The core of the Sun is dense enough that the mutual gravitational attraction of individual hydrogen atoms can provide the necessary kinetic energy for **nuclear fusion** to take place. The temperature of the Sun's core is thought to be around 15 million degrees Kelvin, although at the surface the gas cools to a mere 5700 Kelvin. At the moment, most of the heat generated in the Sun comes from fusion of hydrogen to helium, although evidence shows us that heavier elements are also being made inside the Sun.

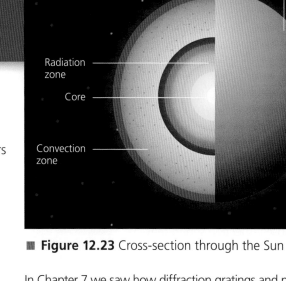

■ **Figure 12.23** Cross-section through the Sun

In Chapter 7 we saw how diffraction gratings and prisms can be used to separate out the wavelengths of light. This technique can be used with astronomical objects too, by fixing a **spectroscope** to the eyepiece of an optical telescope, such that we can obtain the **spectrum** of light from the object. This technique is known as **spectroscopy**. When we analyse the Sun's light in this way, we gain a spectrum showing all the colours of the visible region as we would expect.

■ **Figure 12.24** The solar spectrum

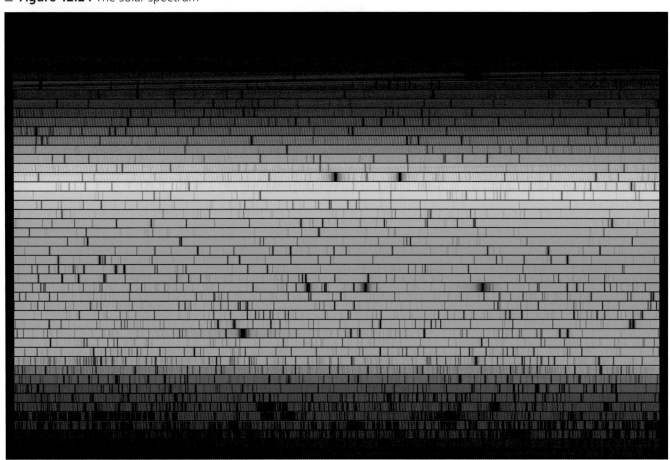

However, close inspection shows that there are dark 'lines' in the spectrum, as though some wavelengths are missing. Astronomers analysed these lines to show that they correspond to the absorption of light by atoms of particular elements.

As we saw in Chapter 11, the Bohr model of the atom suggests that electron orbits are quantized into specific energies. Atoms can absorb, as well as emit, light when their electrons transition between energy states. When an electron 'drops' from a higher energy level to a lower one, it releases the excess energy in the form of electromagnetic energy whose frequency is related to the energy difference between the energy levels by the **Planck–Einstein relationship**

$$\Delta E = hf$$

Where (J) is the energy difference between the electron orbits, f (Hz) is the frequency of the light produced, and h is a constant known as **Planck's constant**,

$$h = 6.63 \times 10^{-34} \, J \, s$$

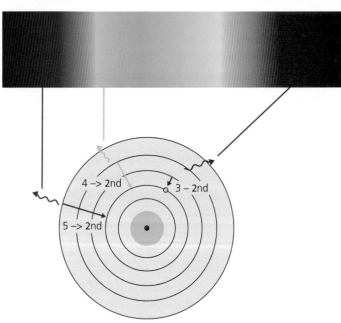

■ **Figure 12.25** Emission spectra and Bohr model

ACTIVITY:
Observing emission spectra

■ ATL

■ **Critical-thinking skills**: Practise observing carefully in order to recognize problems; Test generalizations and conclusions

Aim: To observe emission spectra using a simple spectroscope.

Equipment
- **Small hand-held spectroscope**
- **Fluorescent lighting**
- **Gas discharge tubes**

Write a hypothesis about the kinds of spectra you expect to see from the lighting, and from the discharge tubes. **Outline** the key features you expect to observe.

Method
Safety: Gas discharge tubes use very high voltages and become very hot. The apparatus should be set up by your teacher or technical assistant. Do not touch.

Take the hand-held spectroscope and point the end with the slit at the fluorescent lighting tube. Look through the eyepiece. Carefully observe the spectrum produced. Note any differences in intensity across the spectrum.

Now use the spectroscope to observe the gas discharge tubes. To do this you will need to remove all background light in the room, or place your head and the tube under a dense black cloth. Carefully **observe** the spectrum produced. Note any differences in intensity across the spectrum.

Conclusion
Describe the spectra produced in terms of the frequencies of light visible, any variations in intensity, and **state** the presence of any emission lines. **Outline** the validity of your hypothesis.

Research the expected spectra for the gases present in the tubes and **compare** to your own observations.

◆ Assessment opportunities

In this activity you have practised skills that are assessed using Criterion C: Processing and evaluating.

ACTIVITY:
Analysing emission spectra

■ ATL

■ **Critical-thinking skills**: Interpret data

Look carefully at the spectrum obtained from a sample of hot gases in the laboratory (Figure 12.26).

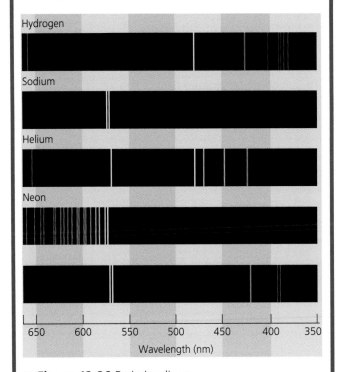

■ **Figure 12.26** Emission lines

State the elements present in the gas mixture that have caused emission lines.

Interpret and **analyse** the spectrum, and so **calculate** the energies of the electron transitions that have caused the emission lines.

◆ Assessment opportunities

In this activity you have practised skills that are assessed using Criterion A: Knowledge and understanding.

How has knowledge of the Universe affected our understanding of our place in it?

OUR PLACE IN THE UNIVERSE

When making observations across the vast distances of deep space, we have to factor in a new variable. In Chapter 7, we saw how the speed of light c in a vacuum is a constant value of around $3 \times 10^8\,\mathrm{m\,s^{-1}}$. The idea that light had a finite speed had been proposed as early as 1676, but accurate measurement of the speed of light was achieved over many, many repeated experiments in the early 20th century CE – particularly by the Polish–American physicist Albert A. Michelson (1858–1931).

The fact that light does not travel at infinite speed and so does not arrive instantaneously has a significant implication. In most circumstances on Earth, the speed of light is so great that the time for light to travel from source to observer is not significant, but when we are looking at distant objects this time lag makes a difference. The light from our own Sun, for example, takes around 8 minutes to travel to the Earth. The light from the next nearest stars takes approximately 4 years to reach us. This means that, when we observe the Sun, we are seeing it as it was 8 minutes ago – and we are constrained to observe even the next nearest stars as they were 4 years previously!

For deep space measurements, astrophysicists define another unit as a 'ruler' for distance, called the **light year**. A light year is **not** a unit of time, but of distance, since

1 light year = distance travelled by light in one year = ct

So 1 light year = 3×10^8 metres per second $\times$ 31 557 600 seconds

$1\,\mathrm{ly} = 9.4607 \times 10^{15}\,\mathrm{m}$

ACTIVITY: The longest ruler

■ ATL

- **Information literacy skills**: Access information to be informed and inform others

In order to gain some sense of astronomical distance, it is worth using the light year to gauge the distances in the observable Universe.

Use **distance light years** followed by the names of the following objects to find their distances.

Object	Distance (ly)
planet Neptune	
Star Proxima Centauri	
centre of our galaxy	
Pulsar CP1919	
Andromeda Galaxy M31	
Hubble telescope deep-space view objects	

Figure 12.27 Hubble deep-space view

Albert Einstein (1879–1955) was the first person to realize that the finite speed of light had some profound implications for our understanding of the Universe. He asked himself a quite simple question: What would the Universe look like if one rode on the back of a beam of light? If you are travelling at the speed of light itself, what could you actually **see**? From this simple thought experiment he derived the **postulates** of the **special theory of relativity**!

Yet, Einstein was not the only person to revolutionize our understanding of our place in the Universe during the early 20th century CE. Another long-debated question was about to gain an unexpected answer: Does the Universe have an end? Isaac Newton for one believed that the Universe was static, infinite, and homogeneous – that is, nothing in it is changing, it goes on forever, and it is the same in every direction. But this view can be challenged with another simple question: Why, then, is the sky dark at night? This question is called **Olber's paradox**.

EXTENSION

Einstein's special relativity, and later his general theory, were to revolutionize physics, our perception of the Universe and our place in it. Find out more about the postulates of special relativity. Why did Einstein call his theory 'relativity'?

THINK-PAIR-SHARE

Think: What answers can you think of for these questions?
- **How many stars are there in an infinite Universe?**
- **How long have they been shining?**

Discuss in pairs: what does Newton's idea of an infinite, static Universe suggest the night sky **should** look like?

Share your ideas in class. Now suggest as many ways as you can think of to explain the fact that the night sky is mostly dark.

In 1928, an astronomer called Edwin Hubble (1889–1953) made an amazing discovery. He was observing the spectra of some different stars, but noticed something surprising. It seemed that the spectrum of light from hydrogen in the stars was different to that seen for hydrogen on Earth.

Hubble thought that he had made a mistake, but when he checked his measurements he realized that the amount by which the wavelengths of hydrogen from stars was 'shifted' (changed) was proportional to the distance of the stars from us. Hubble made a leap of imagination and suggested that this could only be explained by the Doppler effect, which changes the wavelength of waves when an object is moving. Imagine, for example, how the sound of a motor scooter or motorbike changes as it passes.

Hubble realized that light waves could also be Doppler-shifted. The strange thing was that all of the measurements he made suggested the lines were **red-shifted**, or moved towards a longer wavelength. This implied that everything was moving **away** from everything else.

Absorption lines from our Sun

Absorption lines from a supercluster of galaxies. BAS11
$v = 0.07c$. $d = 1$ billion light years

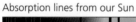 **Figure 12.28** Red-shifted spectra

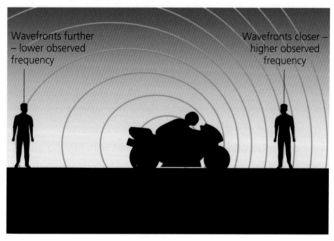

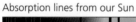 **Figure 12.29** Doppler shifting of sound waves

ACTIVITY: Modelling the Universe

To understand the significance of Hubble's discovery for our understanding of the Universe, take a balloon, and a marker pen. With the balloon deflated, cover it with small dots on both sides. Now inflate the balloon.

Observe what happens to the distances between the dots as the balloon expands.

Describe how the balloon model helps us visualize what is happening in the Universe itself.

Hubble and other scientists deduced that the Universe is expanding. But that, in turn, led to another very strange thought: if the Universe is expanding, then it must have started out very small ... in fact, it must have started out infinitely small!

Hubble measured the red-shifts for many stars and other objects outside our own galaxy. Using other astronomers' estimates of the distances of these objects, he deduced that the speed at which they were moving away from us, or **velocity of recession**, was linearly related to their distance, such that

$$v = Hd$$

Where v is the velocity of recession ($m\,s^{-1}$), d (m) is the distance of the object and H is a constant known as **Hubble's constant**.

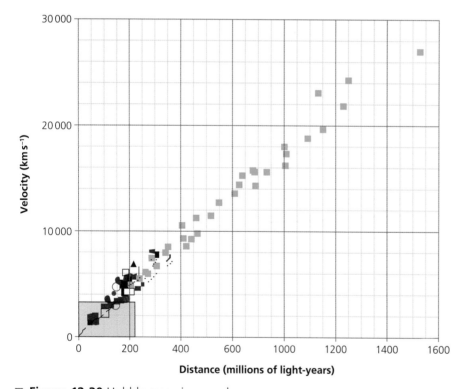

■ **Figure 12.30** Hubble recession graph

THE BIG BANG THEORY

The theory that was elaborated from Hubble's observations is known as the **Big Bang theory**. In the theory, the Universe is neither infinite nor static – rather it began as a point, and has been expanding for an estimated 13.8 billion years. It's important to recognize that the theory does not suggest that all the **matter** in the Universe was squeezed into one place in empty space, rather that all of **space** was only a single point. Just as a balloon expands and stretches as we inflate it, space is itself expanding and stretching, full of matter and energy. At the very beginning, the density of the Universe was unimaginably high, and so was its temperature. The conditions in the first nanoseconds of time were so extreme that matter itself did not exist as we currently know it.

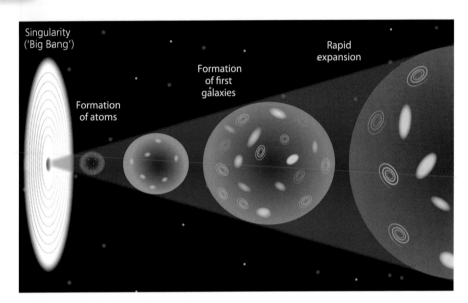

■ **Figure 12.31** Expanding Universe

THINK-PAIR-SHARE

Think: What answers can you think of for these questions?
- **In a Universe of a finite age of 13.8 billion years, what is the furthest distance we can see?**
- **Suggest what this means about the size of the Universe.**
- **Explain your answer.**

Discuss in pairs: How does the expanding Universe theory resolve Olber's paradox?

Share your ideas in class.

EXTENSION

Since Hubble's initial observations, much further evidence has been gathered for the expanding Universe model. **Research** to find out what that evidence is, using, for example, cosmic microwave background.

ACTIVITY: What is the use of space research?

Take action

! Decide what good is space research.

Since the first spacecraft was launched by the Soviet Union in 1957, we have been probing further and further into space with our instruments. Space missions are highly expensive when compared to Earth-based astronomy projects – astronomically expensive, even!

So why do we do it? What do nations gain from space missions?

Research one space mission of your choice.

Write a report on that mission, in which you:

- describe its objectives
- describe how these objectives were achieved using science and technology
- evaluate the mission, comparing its advantages and its disadvantages over Earth-based space research. What were the benefits? What were the costs?

Assessment opportunities

This activity can be assessed using Criterion D: Reflecting on the impacts of science.

■ **Figure 12.32** NASA's *Mars Exploration Rover* Mission (MER)

SOME SUMMATIVE PROBLEMS TO TRY

Use these problems to apply and extend your learning in this chapter. The problems are designed so that you can evaluate your learning at different levels of achievement in Criterion A: Knowledge and understanding.

THIS PROBLEM CAN BE USED TO EVALUATE YOUR LEARNING IN CRITERION A TO LEVEL 3–4

1 In a distant solar system, the planet Mypia has four moons. The table below shows information on the orbits of the moons.

Name of moon	Orbital radius (km)
Petera	250 000
Frankia	380 000
Louon	140 000
Paulon	500 000

   a **Describe** the likely order of the orbital periods of the moons, and **justify** your answer.

     In an experiment to measure the distance to the moons, a small mirror is placed on the moon by a spacecraft. A laser beam is then fired at the mirror and an astrophysicist makes a measurement of the time taken for the laser beam to return to the planet.

   b Make suitable **calculations** to **estimate** the length of time the astrophysicist should expect the laser to take if the distance given for the moon Paulon is accurate.

The astrophysicist finds there are two problems with this measurement technique:
- all the laser beams take a little longer than expected to return to Mypia
- all the laser beams return to a point a small distance to one side of the launch position.
  Assume the speed of light $c = 3.00 \times 10^8 \, \text{m s}^{-1}$

   c **Suggest** two factors that might cause such errors in the experiment.

THIS PROBLEM CAN BE USED TO EVALUATE YOUR LEARNING IN CRITERION A TO LEVEL 5–6

2 The radius of the Earth is 6400 km. A space station is in geostationary orbit at an altitude of 3200 km above the Earth's surface.

   a If the force of gravity experienced by the space station on the Earth's surface is F, **calculate** the force of gravity experienced by the space station in orbit in terms of F.

   b The astronauts on the space station feel as though they have no weight at all, and 'float' inside the space station. **Explain** why this effect occurs.

   c The space station fires its rockets for a short time to increase its speed in a direction tangential to the orbit, but remains in orbit around the Earth. **Describe** the effect of this increase on the orbit of the space station.

THIS PROBLEM CAN BE USED TO EVALUATE YOUR LEARNING IN CRITERION A TO LEVEL 7–8

3 Look carefully at the following experimental data.

Element	Typical emission energies ($\times 10^{-19}$ J)
hydrogen	3.031, 4.092
sodium	3.373, 3.377
oxygen	2.896

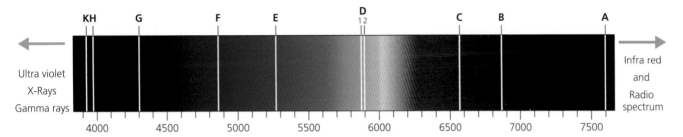

Legend of absorption lines in the visible solar spectrum

Letter	Wavelength (nm)	Colour range
A	759.37	Dark red
B	686.72	Red
C	656.28	Red
D1	589.59	Red orange
D2	589.00	Yellow
E	526.96	Green
F	486.13	Cyan
G	431.42	Blue
H	396.85	Dark violet
K	393.37	Dark violet

■ **Figure 12.33** Absorption lines

The table gives some transition energies for emission lines produced from samples of hot gases in the laboratory.

The second image shows part of the spectrum obtained from an observation of the Sun, taken from an observatory on Earth.

a **Interpret** and **analyse** the diagrams and make suitable **calculations** to identify the elements shown to be present in the spectrum from the Sun.

b The astronomer who made the observation thinks that one of the observed absorption lines is not, in fact, caused by gases in the Sun. **State** which of the elements this is likely to be, and **explain** what might have caused this absorption line to appear in the astronomer's measurements.

The astronomer makes similar observations of the spectrum from a distant star, outside of our solar system. The data she gathers for this star is given below.

Line	Wavelength ($\times 10^{-9}$ m)
C, F	659.5, 489.4
D1, D2	592.9, 593.0
B	686.7

c **Analyse** these new results and **evaluate** any difference from the measurements taken for our own Sun. **Comment** on any differences you note, and **explain** what might have caused them.

Reflection

In this chapter we have zoomed out from our local space neighbourhood, following in the footsteps of the earliest space scientists in their exploration of the Solar System and the objects in it. We have understood the way optical instruments, such as the human eye, lenses and telescopes, are used to manipulate light and achieve magnification of distant objects. We have explored other forms of observational evidence available to astronomers as they looked further out into deep space, such as spectroscopy. We have considered the implications of a finite speed of light, and of the expanding Universe, for our own understanding of our place in space and time.

Use this table to reflect on your own learning in this chapter.		
Questions we asked	Answers we found	Any further questions now?
Factual: What is the scale of the observable Universe and how big are the objects in it? What evidence have we used to elaborate our models of the Universe? What instruments have we used to gather observational evidence? How do stars produce energy?		
Conceptual: Why is the speed of light important to our understanding of the Universe? How do forces shape the Universe?		
Debatable: How has knowledge of the Universe affected our understanding of our place in it? How important is it to know about the Universe beyond our own planet?		

Approaches to learning you used in this chapter	Description – what new skills did you learn?	How well did you master the skills?			
		Novice	Learner	Practitioner	Expert
Information literacy skills					
Critical-thinking skills					
Transfer skills					
Learner profile attribute(s)	Reflect on the importance of being knowledgeable for our learning in this chapter.				
Knowledgeable					

Glossary

absolute zero   Theoretical temperature corresponding to zero internal kinetic energy of an object. Lower fixed point of the Kelvin scale where $0\,K = -273.15°C$.

absorption   Process by which incident energy is converted to internal kinetic energy of an object.

acceleration   Rate of change of velocity.

accommodate   Adjustment of a lens to bring into focus objects at different distances from lens.

aerobic respiration   Biochemical process that releases energy from a sugar in the presence of oxygen.

albedo   Fractional reflectivity of a surface, expressed as a decimal. Albedo = 1 means all incident energy is reflected.

alpha (α) particles   Particle of ionizing radiation, consisting of 2 protons and 2 neutrons, or an energetic helium-4 nucleus.

alternating current (ac)   An electric current that changes direction of flow periodically.

amp-hour   The current used by a load multiplied by the time it is used.

amplitude   Maximum displacement of an oscillation, distance from equilibrium position to peak or trough of a wave.

anthropogenic   Phenomenon created by human activity.

anti-matter   A minority form of matter, thought to have been generated in the Big Bang, which annihilates on contact with matter.

anti-neutrino   A type of charge-less, mass-less, fast particle of anti-matter.

astronomical unit   Mean distance from centre of the Earth to centre of the Sun. 1 au = 149 597 871 km.

atom   Smallest particle of matter comprising an element.

atomic number   Number of an element in the periodic table, corresponding to its proton number (Z).

atomic weight   Mean mass of an element and its isotopes.

Avogadro constant   The number of atoms in 1 mole of any substance. $N_a = 6.02 \times 10^{23}\,mol^{-1}$.

axis   A line of reference.

background radiation   Level of naturally occurring ionizing radiation.

balance   To cancel or be equivalent – in forces, where the resultant of forces equals zero.

bandwidth   A range of frequencies (or wavelengths), such as those passed or absorbed by a filter.

beta (β) radiation   Particle of ionizing radiation consisting of an energetic electron emitted from the decay of a neutron in the nucleus.

biconcave   An object that is concave on both sides, i.e. curving inwards towards its centre point.

biconvex   An object that is convex on both sides – curving outwards towards its centre point.

Big Bang theory   The theory that all the Universe originated in a singularity in space–time and which subsequently expanded.

binding energy   The difference between the sum of the equivalent energies of all the nucleons in a nucleus, and the measured equivalent energy. The energy field associated with the strong nuclear force.

biomass   A mass of material whose origin is biological, such as plant matter.

buoyancy   The upward force exerted by a fluid on a floating object, in opposition to the weight of the object.

carbon dioxide   The molecule CO_2, produced in the combustion of carbon compounds.

carbon footprint   The amount of carbon released into the environment by any process.

carbon monoxide   The molecule CO, produced in the combustion of carbon compounds, especially where combustion is inefficient.

carboniferous period   Geological time period beginning some 360 million years ago and ending some 290 million years ago, during which the Earth was covered extensively in plant life.

carrier wave   A wave that is modulated to carry information.

causation   The phenomenon which links cause and effect.

centripetal force   A force constraining an object to a circular trajectory, acting inwards towards the centre point of the trajectory.

chain reaction   A nuclear reaction which self-sustains as long as fissile material remains.

charge   The quantity which is the origin of the electrical force; bipolar with values that are positive or negative.

ciliary muscles   Muscles in a mammal's eye that stretch the lens to accommodate objects at different distances.

circuit   A closed loop; in electricity, a closed current path from a region of greater positivity to greater negativity.

circuit diagram   A schematic representation of an electrical circuit.

climate-forcing factors   Any factor which is thought to be causing global climatic change.

climate response   Any natural response to a climate-forcing factor.

cloud chamber   A device used to observe the trajectories of ionizing radiations through the condensation of a supersaturated vapour.

combust   Burning; the rapid exothermic oxidation of a compound.

commutator   Device used to maintain the polarity of a coil as it rotates, typically a split ring.

conduction   Heat-transfer process where heat energy is transferred physically between particles as kinetic energy through bonds, typically in solids.

convection   Heat-transfer process where heat energy is transferred physically through the translation or motion of particles, typically in liquids and gasses.

conventional current   An electric current modelled to travel from regions of relative positive charge to relative negative charge.

cornea   A tough clear region over the lens of a mammal's eye which participates in the focussing of light rays.

correlation   A statistical match between two variables.

covalent   A type of atomic bonding where an energetic advantage is gained through the sharing of electrons by the orbitals of two atoms.

critical   Self-sustaining state of a nuclear chain reaction due to at least one neutron being produced per fission.

crumple zone   A region at the front of a vehicle designed to collapse on impact, absorbing kinetic energy.

current   A flow of electrons.

daughter nuclei   The products of a nuclear fission.

decay   A change that leads to a degeneration to a substance of lower mass or energy, such as in a radioactive process.

deceleration   A negative acceleration.

density   The quantity for any substance, in $kg\,m^{-3}$

diminished   An image that is smaller than the object; a decimal magnification.

direct current (dc)   An electric current that flows in one direction only, such as that produced by a battery or cell.

displacement   The net distance between a starting point and an ending point, expressed in terms of directional axes.

distance   The measurement of travel from any given point.

diverge   Spread out, become further apart, such as light rays.

drag   A retarding force produced by the effect of motion through a fluid.

eddy current   A secondary electric current induced inside a conductor by electromagnetic effects.

efficiency   The ratio of useful work obtained to energy put into a process, usually expressed as a percentage.

electrical   Any phenomenon related to the interaction of charge.

electrical grid   The network of equipment used to distribute electrical power over a region.

electromagnetic   Phenomenon caused by the interaction of electrical and magnetic forces.

electromagnetic induction   Generation of an electric current through the interaction of electromagnetic or magnetic fields.

electromagnetic spectrum   The name given to the radiated energy emitted as electromagnetic waves of different frequencies.

electromotive force (emf)   The force produced due to an electromagnetic interaction.

electroweak force   The force which underlies both electromagnetic and weak subatomic force interactions.

element   A pure chemical substance that consists of one kind of atom or its isotopes.

emission   Process by which electromagnetic energy is released from matter.

empiricism   The doctrine that all knowledge is gained through observation of nature.

equilibrium   A balancing point, or resting point, where there is no resultant force.

ethanol   An alcohol produced through the fermentation of biomass.

exothermic   A chemical process in which energy is released as heat.

exponential relationship   A relationship whereby the rate of change of a value is proportional to the value.

eyepiece   Part of an optical instrument containing the lens, which forms a virtual image for observation.

falsification   The doctrine that seeks to test knowledge by finding the conditions under which it is incorrect.

feedback   A quantity that is returned from the output back to the input of a process.

ferromagnetic   Any material that responds to the magnetic field produced by iron.

fissile material   Any material that can undergo nuclear fission.

fission   Process by which nuclei are split into smaller nuclei, releasing energy.

fluid   A substance that can flow or change shape to fill a container.

focal point   Point of convergence of rays in an optical system.

fossil fuel   A carbon-based compound formed in the carboniferous period whose stored energy is released by combustion.

freefall   The state of falling under the pull of gravity, but with sufficient tangential velocity so as to never approach the surface of the gravitational object – to be in orbit.

frequency   The rate of recurrence of a process; the number of oscillations per second.

friction   Retarding force produced by the physical interaction of two surfaces.

fuel   Any substance that contains energy which can be released to do useful work in a process.

fulcrum   A turning point or 'pivot' for a lever.

fuse   Device for limiting electric current in a circuit, usually through melting on resistive heating.

fusion   Process by which smaller nuclei are forcibly combined to produce larger nuclei, with a release of binding energy.

gamma (γ) radiation   Ionizing radiation consisting of very high frequency electromagnetic waves emitted from nuclear vibration.

Geiger–Muller tube   Device that detects ionizing radiations through the change of conductivity of a gas caused by ionization.

geocentric   Model of the Universe which places the Earth at its centre.

gravitational   Effect produced by the action of the gravity force between masses.

gravity   Force of attraction produced by masses.

greenhouse effect   Process of atmospheric absorption of infra-red radiation then re-emission towards the ground, leading to the trapping of heat energy in the atmosphere.

hadron   A relatively massive particle, such as protons and neutrons, consisting of quarks held together by the strong nuclear force.

half-life   The time taken for half the atoms in a substance to decay; alternatively, half the mass.

heliocentric   Model of the Universe which places the Sun at its centre.

heresy   An opinion that opposes or is strongly different to the accepted view, particularly to the views of a Church or religion.

high tension   In electricity, any large potential difference, typically in the kilovolt range.

homopolar motor   A device that produces motion using the interaction of a stationary magnetic field and an electromagnetic field.

Hubble's constant   The constant H relating the recession velocity to the distance of objects in the Universe. H is measured between 60 and $75\,km\,s^{-1}\,Mpc^{-1}$ (kilometres per second per megaparsec).

hypothesis   A provisional theory about a phenomenon, to be tested by measurement and observation.

image   A representation of an object produced through the collection of light rays in an optical instrument.

impulse   The effect of a force for a period of time; impulse = force × time.

induction   An effect caused by electromagnetic interaction, usually the production of an electric current in a conductor.

inertia   A property related to mass describing resistance to change of motion (acceleration).

infra-red   Region of the electromagnetic spectrum which causes increase in the temperature of matter on absorption, held to have wavelengths between 700 nm and 1 mm.

input   Information, energy or matter entering a process.

insulators   Materials which conduct heat or electricity poorly.

intensity   Related to the brightness of an electromagnetic source, or the power transferred per unit area by a wave.

ionization   Removal of electrons from a stable atom to leave an electrically charged particle.

ionizing radiations   Effect of a nuclear decay that causes ionization in matter.

ionosphere   Layer in the Earth's atmosphere consisting of particles ionized by cosmic rays. Reflects radio waves.

irradiance   The power per unit area of electromagnetic radiation that is incident on a surface.

isotope   Atoms with the same proton number but different numbers of neutrons, having the same chemical properties.

Joule   Unit of energy.

kilowatt-hour   Unit of energy used to measure electrical supply.

latitude   Angular distance from the equator.

lens   An optical device used to bring incident light to a real or virtual focus.

light year   Distance travelled by light in one year; $1\,\text{ly} = 9.5 \times 10^{15}\,\text{m}$.

longitudinal oscillation   Oscillation such that the medium vibrates in the same axis as the energy flow.

machine   Any device for doing work.

magnetic   Any phenomenon related to the magnetic aspect of the electromagnetic force.

magnified   When an image is larger than the object.

mass number (M)   Mean of the masses of all isotopes of an element.

measurement uncertainties   Possible deviation from the measured value caused by errors in observation.

medium   A material through which energy can pass.

metallic   A substance with metal properties (malleable, ductile, reflective) and whose atoms bond using delocalized electrons.

microwave   Short-wavelength radio waves in the range $1\,\text{mm}$ to $1\,\text{m}$.

moderator   A material used to slow or absorb neutrons in a fission reactor.

modulation   Combination of two waveforms such that one affects either the amplitude of frequency of the other, usually in order to encode information.

mole   Quantity of a substance containing the Avogadro number of particles.

molecule   A particle containing two or more atoms bonded together.

momentum   Quantity relating to the inertia in motion of an object, equal to mass × velocity.

mutation   A random change in the genetic makeup of an organism. The point at which the human near point eye can first bring an object to focus.

normal force   The reaction force to the weight when it is in the same axis as the weight, that is 'normal' to the surface causing the reaction.

near point   The point at which the human eye can first easily bring an object to focus.

nuclear fission   Process by which nuclei are split into smaller nuclei, releasing energy.

nuclear fusion   Process by which smaller nuclei are forcibly combined to produce larger nuclei, with a release of binding energy.

nucleon number (A)   The total number of protons and neutrons in a nucleus.

objective   Lens used to gather light in an optical instrument.

ohmic   Property of a substance, meaning that it behaves in accordance with Ohm's law $V = IR$ on conduction of electricity.

Olber's paradox   The proposition that an infinite, static Universe ought to be full of light.

optic nerve   Nerve joining the retina in the eye to the brain, transfers information about the image in the eye electrically.

optical fibre   A channel of conduit of higher refractive index than air which is used to transmit light rays by total internal reflection.

orbit   A circular trajectory around any object.

origin   A starting point or initial cause.

oscillation   Any periodic motion around an equilibrium position.

oscilloscope   A device for displaying signal oscillations in real time.

output   Energy, matter or information produced by any process.

parallel circuit   A circuit with multiple, rejoining branches for electric current.

pendulum   A suspended mass used in oscillation.

periodicity   Repeated pattern in chemical properties of the elements.

perturbation   A small motion away from static or dynamic equilibrium.

photosensitive   Property of being responsive to light.

Planck's constant   Proportionality of energy to frequency of a photon, $h = 6.63 \times 10^{-34}\, m^2\, kg\, s^{-1}$.

plasma   State of matter at high energy in which electrons and nuclei are completely dissociated.

positive feedback   Output from a process that, when returning to be added to the input, tends to increase the effect.

postulates   Initial conditions of truth for a theoretical proposition.

potential difference (p.d.)   Difference in the energy due to electrical charge between two points in space or in a circuit.

potential energy   Energy that is stored and available for work.

power   Quantity equal to the rate of doing work, or energy changed in a period of time, measured in Watts.

prism   A transparent object that is used to refract light.

propagate   To travel and spread through a medium.

proton number (Z)   Number of protons in a nucleus.

quantize   To give a precise value to something.

quantum physics   Branch of atomic physics dealing with the consequences of quantized energies in sub-atomic particles; probabilistic model of the subatomic.

quarks   Sub-nuclear particles that combine to form hadrons, such as protons and neutrons, when bonded with strong nuclear force.

radiation   Generally, any energy that is propagated through space. Sometimes used to refer to ionizing radiations specifically.

radioactive   Any substance that is capable of radioactive decay.

radiocarbon dating   Method of obtaining age of a sample of organic material from the radioactive emissions from the carbon-14 atoms it contains.

rationalism   Doctrine that knowledge is most certain when deduced from abstract, logical premises.

ray diagram   Schematic representation using lines to show the direction of propagation of wavefronts.

rays   Lines drawn perpendicular to a wavefront, showing the direction of propagation of an electromagnetic wave.

real current   Electric current modelled in terms of the direction of electron flow, that is from regions of greater negativity to regions of greater positivity.

real image   An image that can be projected as it results from the convergence of rays in real space.

red-shift   Change in observed wavelength of light due to receding motion of an object, especially in space.

reflection   Phenomenon whereby incident light rays are returned back on their path.

refraction   Phenomenon whereby the direction of propagation of light rays is altered, due to change in refractive index, and so wave speed, in the medium.

refractive index   Ratio of the speed of light in a vacuum c to its speed in a given medium.

reliability   Repeatability of a result – extent to which the same result can be achieved through repeated measurement.

renewable   A material or process whose input can be replaced or regenerated.

resistance   Property of a material related to the loss in energy of an electric current on conduction; ratio of voltage over current, measured in ohms, Ω.

resistive heating   The heating effect on a conductor due to increased kinetic energy of its particles through resistance.

resistivity   The resistance per unit length and area of a material.

resistors   Small electronic devices of known resistance used for controlling currents in electronic circuits.

resolution   Ability of an optical instrument to separate two points on a distant object; the angular or physical distance between two points that can just be distinguished on a distant object.

retardation   A negative acceleration.

Sankey diagram   Diagram showing the proportional energetic outputs for a given energetic input to a process.

saturation   Point of maximum absorbance of energy or matter in a material.

scatter graph   Graph showing the relationship between two variables as discreet points.

semiconductor   A material whose electrical properties lie between conductors and insulators; often materials whose conductive properties are manipulated through addition of other materials.

series circuit   A circuit in which loads are in the same circuit path.

slip ring   A device for maintaining constant connection to a rotating coil in a generator.

Solar System   The system of objects for whom the Sun is the principal gravitational centre.

solenoid   An electromagnet arranged as a rod with a central magnetic core.

special theory of relativity   Einstein's theory describing the consequences for motion of an invariant speed of light.

specific energy, specific energy density   The energy content per unit mass of a substance.

spectroscope   A device used to separate component wavelengths from a source of electromagnetic radiation.

spectroscopy   The branch of physics concerned with the analysis of materials through their effects on light.

spectrum   A range of electromagnetic energy ordered by wavelength and frequency.

speed   The rate of change of distance with time.

stepping down   A reduction in supply voltage achieved using a transformer.

stepping up   An increase in supply voltage achieved using a transformer.

strong nuclear force   Force that binds quarks together within hadrons, and that holds nucleons together in the nucleus.

supercritical   State of producing an excess of neutrons per fission in a fission chain reaction.

sustainability   Extent to which a process or energy source can be supported by the Earth's environment.

terminal velocity   The velocity at which an object falling through a fluid has zero resultant force, due to balance of the drag and the weight.

thermonuclear pile   Simple matrix arrangement of fissile material and moderator, intended to produce heat through fission.

thrust   An accelerating force produced typically by an engine.

time period   The period of an oscillation.

Tokamak   Early form of experimental nuclear fusion reactor designed by Soviet Russia.

Torus   Western development of the Tokamak nuclear-fusion design, based on a torus or doughnut-shaped plasma chamber.

transformer   Device for raising or lowering the voltage of an alternating current using electromagnetic induction.

transverse oscillation   An oscillation in which the motion of the medium is at right angles to the direction of propagation of energy.

trend   A pattern with direction in data.

upthrust   A force provided by a fluid that opposes the weight of an object – see **buoyancy**.

validity   The extent to which a measurement supports a hypothesis.

variable   Any quantity whose value in a relationship may change.

velocity   The rate of change of displacement of an object with time.

velocity of recession   The velocity at which distant objects in the Universe are measured to be moving away from us.

virtual image   An image that is not formed by the convergence of rays in real space within an optical instrument, but which can be reconstructed by the eye and the brain.

wave motion   Oscillatory motion in a medium.

wavefront   A line drawn along the peaks of a wave travelling through a medium.

wavelength   Distance between on point in an oscillation and the same point in the next period.

work done   Energy changed by a process, also force × distance moved by the force.

Acknowledgements

The Publishers would like to thank the following for permission to reproduce copyright material. Every effort has been made to trace all copyright holders, but if any have been inadvertently overlooked the Publishers will be pleased to make the necessary arrangements at the first opportunity.

Photo credits

p. 2 *l* © Eléonore H – Fotolia.com; **p.2** *l* © Nasa; **p.3** *l* © md3d – Fotolia.com; **p. 3** *r* © marcel – Fotolia.com; **p.3** *c* © Nasa; **p.5** *l* Public Domain; **p.5** *r* © World History Archive / Alamy; **p.7** *l* Courtesy of BjörnF~commonswiki – Wikipedia Commons (http://creativecommons.org/licenses/by-sa/3.0/); **p.7** *r* © Georgios Kollidas – Fotolia.com; **p.8** *l* © Georgios Kollidas – Fotolia.com; **p.8** *r* © Juulijs – Fotolia.com; **p.9** *l* © charger_v8 – Fotolia.com; **p.9** *c* © James Phelps JR – Fotolia.com; **p.9** *r* © Sergii Moscaliuk – Fotolia.com; **p.11** *t* © Department of Energy's National Center for Electron Microscopy (NCEM) at Lawrence Berkeley National Laboratory; **p.11** *b* © Tomas Abad / Alamy; **p.12** *t l* Courtesy of World Imaging via Wikipedia Commons (Public Domain); **p.12** *t r* Public Domain; **p.12** *b* Public Domain; **p.14** *l* © JOEL AREM/SCIENCE PHOTO LIBRARY; **p.14** *r* © Niels Poulsen mus / Alamy; **p.16** *t* © Vincenzo Lombardo / Getty Images; **p.16** *b l* © The University of Manchester (www.graphene.manchester.ac.uk); **p.16** *b c* © The University of Manchester (www.graphene.manchester.ac.uk); **p.16** *l* © World History Archive / Alamy; **p.18** *t* © odriography – Fotolia.com; **p.18** *b* © passmil198216 – Fotolia.com; **p.22** © http://www.goodfon.su; **p.23** *t l* © CDT Group (http://www.exair.co.uk/site/emissions.html); **p.23** *b l* © Stockbyte via Thinkstock; **p.23** *r* © VRD – Fotolia.com; **p.26** © vvoe – Fotolia.com; **p.27** *t l* © Yi Lu/Viewstock/Corbis; **p.27** *t l* © sciencephotos / Alamy; **p.27** *r* Courtesy of MattWade via Wikipedia Commons (http://creativecommons.org/licenses/by-sa/3.0/); **p.28** *l* © Paul Morris; **p.28** *r* © Paul Morris; **p.30** © Dorling Kindersley/UIG/SCIENCE PHOTO LIBRARY; **p.32** © NASA; **p.33** © NASA; **p.36** © Heiko Küverling – Fotolia.com; **p.40** © caroline letrange – Fotolia.com; **p.41** © Arochau – Fotolia.com; **p.41** *l* © Science and Society / Science and Society; **p.41** *r* © sonya etchison – Fotolia.com; **p.46** Courtesy of Daniel Christensen, Wikipedia (http://creativecommons.org/licenses/by/3.0/); **p.48** © Sharon McTeir; **p.49** © nikkytok – Fotolia.com; **p.50 (1)** © elena_suvorova – Fotolia.com; **p.50 (2)** © K R Robertson, Illinois Natural History Survey; **p.50 (3)** Courtesy of King of Hearts via Wikipedia (http://creativecommons.org/licenses/by-sa/3.0/); **p.50 (4)** © EcoView – Fotolia.com; **p.50 (5)** © David_Steele – Fotolia.com; **p.50 (6)** © Smileus – Fotolia.com; **p.54** *l* Courtesy of Mike Lehmann, Mike Switzerland via Wikipedia Commons (http://creativecommons.org/licenses/by-sa/2.5/); **p.54** *r* Courtesy of Colin via Wikipedia Commons (http://creativecommons.org/licenses/by-sa/4.0/); **p.56-57** © Tom Shaw / Getty Images; **p.58** *t* © Roger Viollet/REX; **p.58** *b* © World History Archive / Alamy; **p.61** © aijohn784 – Fotolia.com; **p.70** Courtesy of Chris Loader (chartingtransport.com); **p.71** Courtesy of Chris Loader (chartingtransport.com); **p.72** *t* Courtesy of Jastrow via Wikipedia Commons (Public Domain); **p.72** *b* Public Domain; **p.73** *t* © ANDREW COWIE / Stringer / Getty Images; **p.73** *b* © NASA; **p.74** *c* Courtesy of Scewing via Wikipedia Commons (Public Domain); **p.74** *t* © DEA / G. NIMATALLAH / Getty Images; **p.79** © KeystoneUSA-ZUMA/REX; **p.81** © Leonid Andronov – iStock Editorial via Thinkstock; **p.82** *t* © NASA; **p.82** *b* Courtesy of Gaius Cornelius / Wikipedia Commons (Public Domain); **p.90** © Pascal Le Segretain / Getty Images via Thinkstock; **p.92** *b* © romti – Fotolia.com; **p.92** *c* © Paul Morris; **p.92** *c* © Sasajo – Fotolia.com; **p.93** *c* © Serghei Platonov – Fotolia.com; **p.93** *t* © myfotolia88 – Fotolia.com; **p.93** *b* © 1996 J. Luke/Photodisc/Getty Images/ World_Religions_35; **p.95** © julie woodhouse / Alamy; **p.95** Public Domain; **p.97** *l* © Sergey Novikov – Fotolia.com; **p.97** *r* © kaimanblu – Fotolia.com; **p.100** *r* © Magdalena Kucova -Fotolia.com; **p.100** *l* © nikkytok – Fotolia.com; **p.103** *t l* © Voyagerix – Fotolia.com; **p.103** *tr* © MARTYN F. CHILLMAID/SCIENCE PHOTO LIBRARY; **p.103** *b l* © NASA/SCIENCE PHOTO LIBRARY; **p.103** *b r* © Professor Ross W. Griffiths FAA Research School of Earth Sciences The Australian National University; **p.104** *t* © William Putman/NASA Goddard Space Flight Center/SCIENCE PHOTO LIBRARY; **p.104** *b* © Ali Kabas / Alamy; **p.109** © SHEILA TERRY/SCIENCE PHOTO LIBRARY; **p.110** © Paul Morris; **p.115** © indigolotos – Fotolia.com; **p.115** © aleksandarfilip – Fotolia.com; **p.115** © Gianfranco Bella – Fotolia.com; **p.115** © AVD – Fotolia.com; **p.115** © fotovapl – Fotolia.com; **p.116** *l* © foto_images – Fotolia.com; **p.116** *r* © kaluginsergiy – Fotolia.com; **p.120** Courtesy of United States Navy via Diego pmc / Wikipedia Commons (Public Domain); **p.123** © BSIP SA / Alamy; **p.123** © EyeOn / Getty Images; **p.127** © sciencephotos / Alamy; **p.136** © Daddy Cool -Fotolia.com; **p.139** © yasar simit – Fotolia.com; **p.144** *l* © inga spence / Alamy; **p.144** *r* Oramstock / Alamy; **p.148** *t* © Tupungato – Fotolia.com; **p.148** *b* Courtesy of Dani 7C3 via Wikipedia Commons (http://creativecommons.org/licenses/by/2.5/); **p.149** *t* Courtesy of Harryzilber via Wikipedia Commons (http://creativecommons.org/licenses/by/1.0/); **p.149** *b* Courtesy of Mike Gonzalez (TheCoffee) via Wikipedia Commons (http://creativecommons.org/licenses/by-sa/3.0/); **p.156** © Max Blatter (www.energie-atlas.ch); **p.158** © Planetek; **p.160** © spumador – Fotolia.com; **p.161** © JohanSwanepoel – Fotolia.com; **p.162** *l* © ClassicStock / Alamy; **p.162** *r* © Clive Streeter / Getty Images; **p.163** © Paul Morris; **p.164** *l* © air – Fotolia.com; **p.168** *t l* © stocksolutions – Fotolia.com; **p.168** *b l* © DiversityStudio – Fotolia.com; **p.168** *r* Courtesy of Threecharlie via Wikipedia Commons (http://creativecommons.org/licenses/by-sa/3.0/deed.en); **p.171** © James Courtney - iStock via Thinkstock; **p.173** © Francisco Javier Gil – Fotolia.com; **p.174** © wdstock / Getty Images; **p.176** *t* © Andriy Solovyov – Fotolia.com; **p.176** *b* © WENDERSON ARAUJO / Stringer / Getty Images; **p.179** © 2013 Antoine Bercovici via Wikimedia Commons (http://creativecommons.org/licenses/by-sa/3.0/deed.en); **p.186-187** ©NASA; **p.187** *r* © Nasa; **p.190** © www.veka-recycling.co.uk; **p.193** Courtesy of Windell H. Oskay, www.evilmadscientist.com via Wikipedia Commons (http://creativecommons.org/licenses/by-sa/3.0/deed.en); **p.198** *t* Courtesy of Marcus Wong via Wikipedia Commons (http://creativecommons.org/licenses/by-sa/3.0/deed.en);

Text credits

Index